THE
NATURAL HISTORY
OF THE BIBLE

A GUIDE FOR BIBLE READERS AND NATURALISTS

THE
NATURAL HISTORY
OF THE BIBLE

A GUIDE FOR BIBLE READERS AND NATURALISTS

PETER
GOODFELLOW

JOHN BEAUFOY PUBLISHING

First published in the United Kingdom in 2017 by John Beaufoy Publishing,
11 Blenheim Court, 316 Woodstock Road, Oxford OX2 7NS, England
www.johnbeaufoy.com

10 9 8 7 6 5 4 3 2 1

ISBN 978-1-909612-98-3

Printed and bound in Malaysia by Times Offset (M) Sdn. Bhd.

Edited, designed and typeset by Gulmohur Press
Project management by Rosemary Wilkinson
Cartography by William Smuts

PAGE 2: TRISTRAM'S GRACKLE; PAGE 3: DONKEY

CONTENTS

CHAPTER I

THE PHYSICAL GEOGRAPHY AND NATURAL VEGETATION OF THE HOLY LAND

When Moses led the Israelites out of Egypt they wandered in the desert for 40 years until they came within sight of the land that God had promised them. Here God told Moses he could not cross the river to enter The Promised Land; instead he led him to the 'top of Pisgah [to] look west and north and south and east (Deuteronomy ch. 3, v. 27). Pisgah, or Mount Nebo, is in the northern part of the land known as Moab, on the north-eastern corner of the Dead Sea. Here Moses could see all the land his people would inherit, and from here we can get a sense of the wide variety of countryside that the wildlife of the Bible live in: mountains, forest, farmland, water, desert and human habitation.

The land we are reading about does not fit neatly into a modern political map. Although for many years the Jews lived in Egypt and Babylon, most of their story takes place in 'Palestine', and this is now covered by two modern nations, Israel and Jordan. The politics of the area today mean that many people are sensitive about who has the right to various parts of this land, so in the pages that follow 'Palestine' means the land in the Bible story; 'Jordan' refers to the river or the valley which bears its name, and 'Israel' refers to the ancient name of the Jewish kingdom which lay north of Jerusalem, as opposed to Judah which lay to the south, unless it is clearly stated otherwise.

It's the mountains which shape the lands of the Bible, the Holy Land as Christians call it, that lie along the eastern shore of the Mediterranean Sea. Near the coast two parallel ranges of mountains run from north to south. The western range is unbroken from Mount Taurus (3,685 m/12,089 ft asl) in Asia Minor (in modern Turkey) in the Taurus Mountains overlooking Tarsus where St Paul was born, to Qornet es Saruda (3,087 m/10,127 ft asl) in Lebanon, which is capped in perpetual snow. Thence the uplands go into the districts north and south of Galilee, rising to Mount Carmel by the coast (actually part of a ridge 39 km/24 m long, 546 m/1,791 ft asl at its highest point), and the plateau-like hill country which includes Jerusalem (754 m/2,474 ft), Bethlehem (775 m/2,543 ft), Hebron (3,050 ft/930 m) and most of the historical sites of Palestine. South of here is the Negev, 'the wilderness' as recorded, for example, in 'The voice of one crying in the wilderness' (Isaiah ch. 40, v. 3 and St John ch. 1, v.23), or 'the desert' as in modern translations.

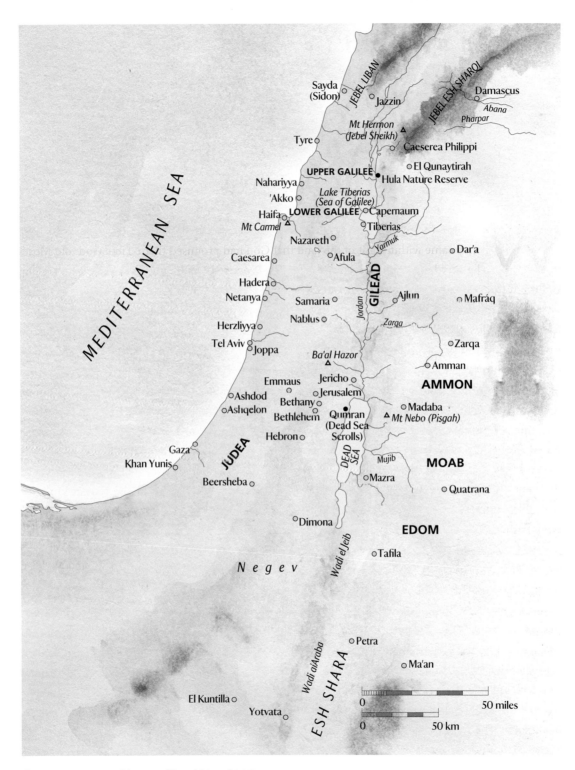

MEDITERRANEAN SEA

JEBEL LIBAN

JEBEL ESH SHARQI

Sayda
(Sidon)

Jazzin

Damascus

Abana

Pharpar

*Mt Hermon
(Jebel Sheikh)*

Tyre

Caeserea Philippi

UPPER GALILEE

El Qunaytirah

Hula Nature Reserve

Nahariyya

'Akko

*Lake Tiberias
(Sea of Galilee)*

Haifa

LOWER GALILEE

Capernaum

Mt Carmel

Tiberias

Nazareth

Yarmuk

Dar'a

Caesarea

Afula

GILEAD

Hadera

Ajlun

Mafráq

Netanya

Samaria

Jordan

Herzliyya

Nablus

Zarqa

Tel Aviv

Zarqa

Joppa

Ba'al Hazor

Amman

Emmaus

Jericho

AMMON

Ashdod

Jerusalem

Bethany

Madaba

Ashqelon

Bethlehem

Quimran
(Dead Sea
Scrolls)

Mt Nebo (Pisgah)

Hebron

Gaza

JUDEA

Mujib

MOAB

Khan Yunis

*DEAD
SEA*

Mazra

Beersheba

Quatrana

Dimona

EDOM

N e g e v

Wadi el Jeib

Tafila

Wadi al Araba

Petra

ESH SHARA

Ma'an

El Kuntilla

Yotvata

0 ———————— 50 miles

0 ———————— 50 km

Topographical Map of The Holy Land

The eastern range which runs from north to south through Syria is generally less high than the western range, but it is the source of the great rivers of the area – the Orontes, the Pharpar and the Jordan. Its highest point is snow-capped Mount Hermon or Jebel Sheikh, i.e. the Prince Mountain (2,814 m/ 9,232 ft asl), which looks down on the whole of Palestine, and is thought to be the site of the Transfiguration of Jesus (St Matthew ch. 17, v. 1). The hills go on through Gilead, Moab and Edom on the east side of the Jordan River and the Dead Sea, including Mount Nebo where Moses stood.

Between the two parallel chains is the country's most extraordinary physical feature, the deep valley or trench which holds the River Jordan, Sea of Galilee, and the Dead Sea, all part of what is still commonly called The Great Rift Valley, which runs from the Beqaa Valley in Lebanon to Mozambique in eastern Africa, a total of c. 6,000 km/3,700 miles. The Dead Sea or Salt Sea 429 m/1,047 ft *below sea level*, is the lowest elevation on earth, and is one of the saltiest bodies of water in the world, which makes it a very harsh environment to live in, hence its name.

Great upheavals in the earth's crust helped create Palestine, and that activity is still not finished. Earthquakes are mentioned several times in the Bible and there are strikingly precise references, such as the prophet Amos saying,

⮞ AMOS
CH. 1, V. 1

... what he saw concerning Israel two years before the earthquake, when Uzziah was king of Judah and Jeroboam was king of Israel. ⮞

PALESTINE

This was clearly a major shock, long remembered. Two of the best known stories in the New Testament record the moment Jesus died on the cross and

 ST MATTHEW
CH. 27, VV. 51-52

At that moment the curtain of the temple was torn in two from top to bottom. The earth shook and the rocks split . The tombs broke open ...

and the time St Paul and Silas were in prison in Philippi in Macedonia:

ACTS OF THE APOSTLES
CH. 16, VV. 25-26

About midnight Paul and Silas were praying and singing hymns to God, and the other prisoners were listening to them. Suddenly there was such a violent earthquake that the foundations of the prison were shaken.

No-one was hurt, no-one escaped, and Paul and Silas calmed the jailer's fears so that he became a Believer.

Modern Israel regularly has earth tremors and has had several earthquakes in recent years which shook buildings but caused little damage and caused no casualties.

The Holy Land has what geographers call a Mediterranean climate characterized by long, hot, dry summers and short, cool, rainy winters, as modified locally by altitude and latitude. The climate is determined by its location between the subtropical aridity characteristic of Egypt and the subtropical humidity of the eastern Mediterranean area. January is the coldest month, with temperatures ranging from 5°C/41°F to 10°C/50°F, and August is the hottest month at 18°C/64°F to 38°C/100°F. There is a marked, dry season, which is extraordinarily regular, starting on or about 15 June and ending about 15 September. During this time day after day is the same. Winds develop which can be quite strong at times, but are usually steady, so this is the dependable period in biblical times for shipping; winds in the rainy season are much more changeable and dangerous as when St Paul was being taken to Rome (Acts ch. 27), or when Jesus calmed the storm on Galilee (St Mark ch. 4 and St Luke ch. 8). The cooler rainy season lasts for the rest of the year. The winter rain is not so regular, and appreciable rainfall may not occur till after Christmas. Average temperatures in the Jordan valley from May to October are usually well over 30°C/86°F; winters are much cooler, below 10°C/50°F in December to February; in Jerusalem the average August temperature is 29°C/85°F, whereas in January it is only 8°C/47°F. Frost may occur then at night, and snow may even fall in the south, as happened in the winter of 1991–1992 when banana plantations in Jericho were ruined. The temperature and rainfall vary greatly through the area because of the landscape; the western-facing slopes are wetter than the eastern, and temperatures tend to increase the further one goes from the sea, and as one goes further south.

Rockrose

As far as we know Palestine has always had a coastline, indeed for millions of years the area was under the sea, which has resulted in much of the land now being formed of the maritime deposits of limestone, chalk and sandstone. The wide variety of physical features, from snow-topped mountains to fertile plains to deserts, means natural vegetation of the land is very different from one area to another. The Psalmist many years ago wrote 'the earth is full of your creatures' (Psalm 104, v. 24). This was certainly true in biblical times, when the landscape gave a great variety of habitats, even though the underlying rock was largely limestone or sandstone, except for the alluvial plain and Rift Valley.

The lowlands behind the sand dunes of the coast are now orange groves, vineyards, and fields of cereals and vegetables, but in ancient times there would have been large areas of marshland as well as settlements. The hill country still has extensive areas of wild flowers in spring and shrubland of Dwarf Oaks, Carob trees and Rockrose. This scrub can be very dense, particularly around Mount Carmel and on the hills around Galilee. The Rift Valley has vegetation the thickness of jungle and tropical heat in summer to go with it! The heat and the water create a lush, green ribbon of vegetation, which closely follows the river and consists of Tamarisk, and Christ's Thorn which can grow to a height of 15 m/over 45 ft. Many water birds are still found north of Galilee at the Hula (or Huleh) nature reserve. It is a marsh of papyrus and other plants, but is only a fraction of its ancient size because much has been drained for farmland. The Negev desert in the south is never completely bare and sandy as one often thinks a desert is; the hills may be bare and stony but many wadis (valleys which may have a stream in the rainy season) and hollows have stunted bushes, and a good rainy season will bring a flush of grass and flowers.

In summary, the climate, landforms and human habitation through the centuries have resulted in there being several different plant communities in the region:
- scrub and shrublands called *garigue* and *maquis* respectively. The former has many soft-leaved, ground layer plants, such as Lavender, Cistus, Senecio, Rosemary, thistles and Wild Thyme; the latter is a shrubland biome typically consisting of densely growing evergreens, such as Holm Oak, Kermes Oak, Tree Heath, Strawberry Tree, Sage, Juniper, Buckthorn, Spurge Olive and Myrtle.
- grasslands
- woodlands of oak and pine, and forests, which are distinct from woods by having a closed canopy, of mostly evergreen oaks and pines.

DWARF OAK

CHAPTER 11

TREES AND SHRUBS

In the Bible's story of Creation there is one careful, simple description of botanical life on earth:

⌇ GENESIS
CH. 1, vv. 11-13

Then God said, "Let the land produce vegetation: seed bearing plants and trees on the land that bear fruit with seed in it, according to their various kinds. And it was so. The land produced vegetation: plants bearing seed according to their kinds and trees bearing fruit with seed in it according to their kinds. And God saw that it was good. And there was evening, and there was morning – the third day. ⌇

The author wrote this maybe 4,000 years ago, part of an explanation of the creation of the world. The creation story of the appearance in steady order of fish, plants, animals and birds in chapters 1 and 2 of Genesis does not prevent many God-fearing people today from believing in science and evolution, and understanding the explanations of climatologists, geographers, zoologists and botanists. They describe the Mediterranean basin as one region of five in the world to have a distinct climate (see p. 9) and the effect this has on the flora. The botany of the Holy Land is indeed richly varied, from the wilderness of Beersheba in the south to the top of Mount Hermon in the north. Scripture has many references to this, and modern travellers still remark, especially during pilgrimage time at Easter, on the beauty of the spring flowers.

About 3,000 species of Palestinian flora are known today, but the Bible records only about 130, mostly those that are useful or ornamental. Few species can be identified with certainty; a large proportion of the names are generic, such as nettles, briars and grass. It may seem frustrating to some readers, who are familiar with the detailed classification of plants to be found in modern fieldguides, to learn that the Bible refers simply to (1) *deshe*, signifying all low plants, (2) *'esebh*, including herbaceous plants; and (3) *'es peri*, embracing all trees (Genesis ch. 1, vv. 11–12).

The Bible records trees repeatedly to describe the richness of Palestine's landscape, and to illustrate figuratively the Jews' life in war and peace. Trees are part of the people's spiritual understanding of God's creation:

⌇ ISAIAH
CH. 44, v. 23

Burst into song, you mountains, you forests and all your trees … ⌇

Most telling of all is the Psalmist's assessment of Man whose nature is compared with the characteristics of trees; his ability to do good things is likened to the palm, and his being big and strong to the Cedar:

The righteous will flourish like a palm tree,
they will grow like a cedar of Lebanon;
planted in the house of the Lord,

~ PSALM 92, vv. 12-15

they will flourish in the courts of our God.
They will still bear fruit in old age,
they will stay fresh and green,
proclaiming, "The Lord is upright;
he is my rock, and there is no wickedness in him." ~

CEDAR

It can fairly be argued, however, that neither of these trees can claim to be first in the people's hearts and minds. There is an ancient tradition still practised by people in many countries of there being a special 'meeting tree' where the village elders gather and where special holy rites are conducted, as I have seen in The Gambia, West Africa and on the Salt Trail in the foothills of the Himalayas in Nepal.

Our story of 'all kinds of trees' (Genesis ch. 2, v. 9) really begins with the wanderings of Abram and the Israelites after Abram (only much later did God rename him *Abraham*) was told he was to leave his homeland in Mesopotamia in the 17th century BC and go west to a new land:

The Lord had said to Abram, "Leave your country, your
people and your father's household and go to the land
I will show you."....
So Abram, left as the Lord had told him.... and they set

~ GENESIS CH. 12, vv. 1-7

out for the land of Canaan, and they arrived there.
Abram travelled through the land as far as the site
of the great tree of Moreh at Shechem ...
The Lord appeared to Abram and said, "To your offspring
I will give this land." So he built an altar there to the
Lord, who had appeared to him. ~

A famous sanctuary already existed at Shechem, which is on the edge of the Plain of Philistia to the west of the Dead Sea. A great tree was there, declaring this was a holy place, so it was a natural place for God-fearing Abram to pitch camp. It is also thought that he camped there so that his own large herds of sheep, cattle and goats did not damage the cultivated land of the Canaanites. We have then perhaps the first written reference to mankind being thoughtful about his relationship with the environment, the study of which we call *ecology*. Abram's animals could graze under the trees in the wooded foothills. This holy site is also mentioned later in the beginning of Genesis chapter 35:

> *Then God said to Jacob* [Abram's grandson], *"Go up to Bethel and settle there, and build an altar there to God who appeared to you when you were fleeing from your brother Esau."*

So Jacob settles by the oak at Shechem, and calls the place El Bethel, which means God of Bethel. The prophet Amos spoke in about 750 BC of the strength of the oak as a metaphor for that of men but that strength could not successfully fight against the Israelites with God on their side:

AMOS
CH. 2, V. 9

> *I destroyed the Amorite before them,*
> *though he was tall as the cedars*
> *and strong as the oaks.*
> *I destroyed his fruit above*
> *and his roots below.*

In other words the Amorites (another name for the inhabitants of Canaan) were totally defeated.

In modern Israel, there are several species of oak, and unfortunately there are five Hebrew words in the Bible that are indiscriminately translated as 'oak': *'ayl, 'elah, 'elon, 'allah* and *'allon*. From Isaiah ch. 6, v. 13 it appears that the *'elah* is different from the *'allon*; in fact, *'ayl, 'elah* and *'elon,* are understood by some to be the terebinth, and *'allah* and *'allon* to represent the oak. The best-known genus for the oak is *Quercus*. It is represented in Palestine by seven species, of which three species are common in Palestine.

The most widespread was the Prickly Evergreen Oak *Quercus coccifera,* also known as the Syrian or Kermes Oak. It is native to the Mediterranean region, is an evergreen scrub oak, can grow up to 5 m (16 ft) tall, but is rarely taller than 2 m (6 ft), and has stiff spiny leaves. One 19th century writer, Dr Hooker, wrote, 'It covers the hills of Palestine with a dense brushwood of trees … thickly covered with small evergreen rigid leaves, and bearing acorns copiously'. Its scientific name *coccifera* may immediately make the reader think of the food colouring, cochineal. The tree is an important food plant for a Kermes Scale insect *Porphyrophora/Kermes/Coccus* spp., one of a group of insects from the females of which a crimson red dye is obtained. The dye was known as early as the 8th century BC. As many as 20,000 insects were needed to produce about 300 g (10 oz) of the dye. The name *Canaan* actually means *Land of Purple* because of the importance to the economy of the dyeing industry. Today the tree's distribution is much reduced by the spread of the larger (up to 25 m/80 ft tall) Evergreen or Holm Oak *Quercus ilex,* and by deforestation over the centuries for the production of charcoal and for agricultural ground. Kermes Oak is a hardy tree, able to flourish on rough, hilly ground, saplings from its acorns grow easily, and it can withstand quite heavy grazing. Abram settled well!

There is also 'a great tree', mentioned in Judges, in the 10th century BC:

≫ JUDGES
CH. 9, V. 6

*Then all the citizens of Shechem and Beth Millo
gathered beside the great tree at the pillar in
Shechem to crown Abimelech king.* ≫

The association between individual trees and people has a long history. The oak, for example, features in the European mythology of the Greeks where it was the tree sacred to Zeus, the king of the gods; and also in Baltic, Celtic and Norse mythologies. In Europe, the Roman writer, Pliny the Elder (23–79 AD) wrote that the veneration of trees was a universal custom. In North America giant Californian Redwoods *Sequoia* sp. were revered (and still are, if the number of visiting tourists is anything to go by). There the famous writer about the environment, Henry David Thoreau, declared in the mid 1850s that mankind needed forests 'for inspiration and our own true recreation'. In England the association was recorded in Thomas Hardy's novel *The Woodlanders* (published 1887). In the present day The Woodland Trust and The National Trust in the U.K. have a plan called The Ancient Tree Hunt to make a database of all its great and ancient trees; so far well over 100,000 have been uploaded, a very high proportion of which are oaks, and many more are waiting to be dealt with. Abram had this rich feeling for the sacredness of land and what grows there.

HOLM OAK

Abram separated from his nephew Lot. Lot chose the Plain of Jordan near Sodom, Abram went in the opposite direction to the hills of Hebron, and settled among 'the great trees of Mamre' (Genesis ch. 13, v. 18). It is not surprising that Abram chose the land of the oaks to settle in. This was very near the holy site at Shechem. Once again he prevented what might have been environmental damage and tribal strife by settling away from existing settlements. He almost certainly was in well-wooded oak habitat.

A much less happy tale of an oak is in the story of the death of David's friend, Absalom. David had amassed a Judean army to fight against Israel in whose army was Absalom. David had appointed three commanders, one of whom was Joab, and instructed them not to harm his friend.

2 SAMUEL
CH. 18, VV. 9-10

Now Absalom happened to meet David's men. He was riding his mule, and as the mule went under the thick branches of a large oak, Absalom's hair got caught in the tree. He was left hanging in midair, while the mule he was riding kept on going. When one of the men saw this, he told Joab, "I have just seen Absalom hanging in an oak tree."

Joab angrily told the soldier he should have killed him and received a reward of gold. Despite the soldier's reminding him of David's instruction, Joab took three javelins, and with ten of his armour bearers, went and killed Absalom, threw him in a big pit in the forest and covered the body with rocks. David's sorrow and the aftermath of the battle are told in detail in verses 18 and 19 of the same chapter.

Towards the end of Joshua's life, in the late 1300s BC, he assembled all the tribes of Israel together again at Shechem and there they witnessed that

JOSHUA
CH. 24, VV. 24-26

*"We will serve the Lord our God and obey him."
On that day Joshua made a covenant for the people and there at Shechem he drew up for them decrees and laws. And Joshua recorded these things in the Book of the Law of God* [now preserved in the book Deuteronomy]. *Then he took a large stone and set it up there under the oak near the holy place of the Lord.*

That was the seventh memorial that the Israelites had erected to remind them of what the Lord had done for them through faithful leaders such as Moses and Joshua, even though they had several times forsaken the Lord and worshipped other gods. Seven was considered the number of completeness, so this memorial under the sacred tree at Shechem was especially holy for the Israelites.

During all this time of wandering for 40 years in the desert led by Moses, and the following years in Canaan led by Joshua, the Israelites had carried the tablets of the Ten Commandments in the Ark of the Covenant. When Moses was on Mount Sinai talking to God:

〜 EXODUS
CH. 25, VV. 1-16

The Lord said to Moses, "Tell the Israelites to bring me an offering ... Then have them make a chest of acacia wood ... and poles of acacia wood to insert on the sides of the chest to carry it. Then put in the Ark the Testimony [i.e. the ten commandments] *which I will give you.* 〜

ACACIA

Moses said that the Lord had chosen Bezalel to be chief craftsman in command of the skilled men who were to make the travelling temple, the Tabernacle, and the Ark:

> *Bezalel made the Ark of acacia wood – two and a half cubits long,*
> *a cubit wide, and a cubit and a half high* [i.e. about 1.1m /3¾ ft
> long x 0.7 m/2¼ ft wide and high].

The Ark was designed to be carried, fixed on the base to two long acacia poles. Interestingly this shrine compares closely with a roughly contemporary shrine found in the tomb of King Tutankhamun of Egypt (died c.1350 BC). This sort of shrine Moses would surely have seen during his time growing up in Egypt. But the Ark and the supports to the Tabernacle were not the only things of wood. The furnishings in the Holy of Holies, the table, the altar of incense and the altar of burnt offerings were also made of acacia.

≈ EXODUS
CH. 39, V. 32, 42-43

> *So all the work on the Tabernacle, The Tent of Meeting, was*
> *completed … The Israelites had done all the work just as the*
> *Lord had commanded Moses. So Moses blessed them … and*
> *the glory of the Lord filled the Tabernacle.* ≈

There are hundreds of species of acacia, related to mimosas, across the world, growing in tropical or warm climates. Most of those in the Middle East are not tall trees like the oaks but are hardy, thorny bushes or shrubs just a few metres tall, which are a distinctive feature of semi-desert.

As they wandered after they left Egypt, the most common tree the Israelites would have seen would have been the acacia. It has a hard, enduring, close-grained wood, which is avoided by wood-eating insects so it makes good furniture. Emmanuel Swedenborg (1688–1722), the Swedish scientist and theologian, wrote that it 'denotes the good of merit and of justice, which is of the Lord alone'. That is a fitting way to describe the use to which the Israelites put it. *Acacia seyal* is thought to be the species known in Hebrew as the *shittah-tree,* which supplied the *shittim-wood* as recorded in the Bible in Exodus chh. 25–30. The Jews used it for building the Tabernacle and the Ark. Canon Henry Tristram wrote in his book, *Flora and Fauna of Palestine,* that 'There can be

GUM ARABIC TREE

no question as to the identity of the Shittah with the acacia, the only timber tree of any size in the Arabian desert'. It flourishes in the driest situations, scattered over the Sinai peninsula, and in the ravines which open into the Dead Sea.

Acacia seyal is perhaps best known today for its commercial value. It is one of two species of acacia that produce gum arabic. This is used in medicines, the food industry, by artists as a paint thickener, in the production of ceramics and in lithography (the way many of this book's pictures would originally have been printed). The 'burning bush' that Moses saw (Exodus, ch. 3, v. 2) was most likely an acacia miraculously lit by the sun. The prophet Joel in maybe the 9th century BC passed on a blessing from God to the Israelites:

JOEL
CH. 3, V. 18

In that day the mountains will drip new wine,
and the hills will flow with milk;
all the ravines of Judah will run with water.
A fountain will flow out of the Lord's house
and will water the valley of acacias.

The prophet Isaiah wrote poetic words of help and hope from God to the Israelites who were in exile in Babylon in the 6th century BC. The acacia is one of several trees mentioned to show how richly fertile the land will become:

ISAIAH
CH. 41, VV. 18-19

I will turn the desert into pools of water,
and the parched ground into springs.
I will put in the desert
the cedar and the acacia, the myrtle and the olive.
I will set pines in the wasteland,
the fir and the cypress together,
so that people may see and know,
may consider and understand,
that the hand of the lord has done this, the
Holy One of Israel has created it.

Earlier, in verses 5–8 the prophet had described how the wooded landscape changed, firstly deforested by felling, then being restored naturally. For centuries the kings of Babylon and Assyria had sent woodsmen to fell and take away cedars, *Cedrus libani* in particular, because they were greatly prized timbers in the construction of buildings. It was no surprise therefore, that when King Solomon wanted to build a temple, he no doubt remembered that his father King David had built a palace with the help of cedar logs and carpenters sent by King Hiram of Tyre (2 Samuel, ch. 5).

CYPRESS

1 KINGS
CH. 5, V. 9

So Hiram sent word to Solomon: "I have received the message you sent me and will do all you want in providing the cedar and pine logs. My men will haul them down from Lebanon to the sea, and I will float them in rafts by sea to the place you specify."

The building of Solomon's temple was begun in the fourth year of his reign, c. 966 BC and took about seven years to complete, employing thousands of men, mostly non-Israelite conscripts, who were organized by Adoniram who had done the same responsible work for King David. The temple's measurements are carefully recorded in 1 Kings ch. 5 v. 8 and 2 Chronicles ch. 1, v. 7: 60 cubits long, 20 cubits wide and 30 high, i.e. about 27 x 9 x 13.5 m (90 x 30 x 45 ft). The temple was roofed with beams and cedar planks; beams of cedar attached side rooms to the building; and the interior walls of the temple were lined from floor to ceiling with cedar boards, so that no stone could be seen. The floor was covered with pine planks, which were most probably from the Cypress *Cupressus sempervirens*, the third most valuable tree at this time, alongside the oak and the cedar. Another contender for the wood could be the Aleppo Pine *Pinus halepensis,* which is native to the whole Mediterranean region. It is a big tree, up to 25 m (82 ft) tall. Isaiah quotes God's promise that He 'will set pines in the wasteland, the fir and the cypress together' (Isaiah ch. 41, v. 19). It is valued for its fine, hard timber. There is doubt among scholars about the translation of the word which becomes 'pine'; 'fir' is more secure, and Canon Henry Tristram in his book says the Aleppo Pine 'is especially the Fir Tree of Scripture, and is only inferior to the Cedar in size'. In English today we still use 'fir' and 'pine' indiscriminately.

The Holy of Holies was built within the main building, also using much cedar wood, but having the five-sided door jambs and the doors themselves made of olive wood (described later). On the walls around the temple, palm trees and open flowers were carved, which made the whole place seem like a recreation of The Garden of Eden. Two large cherubim, carved from olive wood and covered with gold, embellished the inner sanctuary, and stood as guardians by the Ark.

Solomon also built a huge Palace of the Forest of Lebanon for himself (1 Kings ch. 7). Although the scriptures note precisely the number of items in the temple's furnishings, we are not told how much timber was felled and used. The deforestation caused just to complete this project must have been considerable, and appears to go counter to the King's desire to decorate the interior with splendid carvings of God's creation.

Time, along with the exploitation of the Cedar's wood, has led to a decrease in the number of Cedar trees in Lebanon. However Lebanon is still known for its Cedars, as they are the emblem of the country and the symbol on the Lebanese flag. The trees survive in mountainous areas, where they are the dominant tree species. This is the case on the slopes of Mount Makmel that tower over the Kadisha valley, where the Cedars of God are found at an altitude of more than 2,000 m (6,600 ft). These trees have reached a height of 40 m (130 ft), and their trunks may be 2.5 m (over 8 ft) in diameter.

Was Solomon's desire to build the best he could imagine for God's house any different from that of the godly men in Europe who built the enormous Christian cathedrals of Canterbury, Notre Dame and St Peter's? Still today mankind builds huge temples or churches, because that seems the best way he can show his devotion to God. The National Cathedral in Brasilia, Brazil, the extraordinary La Sagrada Familia cathedral in Barcelona and the Crystal Cathedral of glass in Garden Grove, California, USA, are just three spectacular, modern examples.

PINE NEEDLES
AND CONE

Most Jewish holidays show a close link between events in their history and the agricultural life they lived. One such festival, 'The Festival of Booths or Tabernacles' known by them as Sukkot, is still celebrated by the faithful for seven days from the 15th day of the seventh month of Tishri (part of our September/October). It is also known as the Festival of the Ingathering; in other words it is a harvest festival, celebrating the bringing in of all the fruit and crops to storehouses. This is one of the Israelites' three Pilgrimage Festivals (together with Passover or the Feast of Unleavened Bread to remember the escape from Egypt, and the Festival of Weeks or Harvest, to celebrate the first barley harvest [see also chapter 3]).

God said to Moses:

LEVITICUS
CH. 23, vv. 40-43

On the first day you are to take choice fruit from the trees, and palm fronds, leafy branches and poplars [or willows], and rejoice before the Lord your God for seven days ... All native born Israelites are to live in booths so that your descendants will know that I made the Israelites live in booths when I brought them out of Egypt.

Still today, orthodox Jews in particular will make a booth from willow and palm fronds, decorated with myrtle, and with an offering of fruit (see p. 27 for details of this traditional offering, the Citron *Citrus medica*). These are samples of the four trees that God names (Willow or Poplar, Date Palm, Myrtle and Citron).

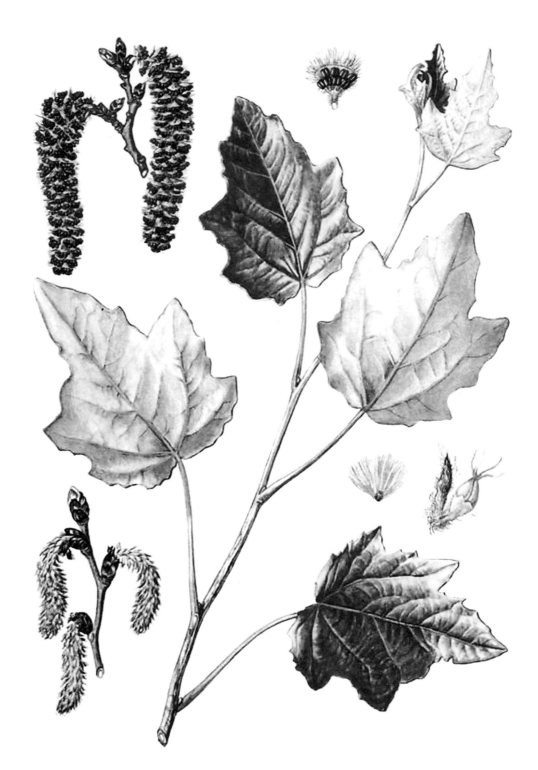

Poplar Leaves and Catkins

When the Israelites came to the banks of the Jordan they found two abundant trees. There was the tree translated into English as the poplar *Populus* sp. or in older translations as the willow *Salix* sp.; and the Oleander *Nerium oleander*, which Canon Tristram said 'lines every wadi from Dan to Beersheba, and which beyond every other shrub in the country must rivet the attention of the most unobservant traveller' because of its gorgeous flowers. The willow is so dependant on water that it became a symbol of the Jews' prayer beseeching God to save them from drought by bringing the winter rains, which would enable them to prepare the land for next year's crops. It is natural that the willow became a necessary part of The Festival of Tabernacles. Several species of poplar are found along the River Jordan and two naturally can claim to be the right translation of the Hebrew word here, *libueh*, which basically means 'white'; they are the White Poplar *Populus alba* and the Euphrates Poplar *P. euphratica*, both of which have white, silvery undersides to their leaves. The Oleander

DATE PALM

is not recorded in the Bible. Was that because the Hebrews knew it is poisonous in all its parts and so considered it 'unclean'? On the other hand, is it the 'rose bushes in Jericho', or 'a rose growing by a stream of water' (Ecclesiasticus ch. 24, v. 14 and 39, v. 13)? Oleanders grow by streams, have beautiful pink and white blossoms, and many a tourist besides myself will have enthused at the sight of them.

The second tree in the list, the Date Palm *Phoenix dactylifera*, probably originated in the Middle East, but it is so widely cultivated now that its original wild state is unknown. It soon became very important in the life of the Israelites as they wandered the desert for 40 years, because the palms also need water, so they are a sign of an oasis, a safe place to camp. The palm is thought of to this day in remembrance of the booths built in the desert. The people soon learned that they not only provide fronds to shelter under and for building, but also high energy food, camel fodder and fibres for weaving baskets and rope.

To those who live in the Middle East, the palm tree is probably as important as or more important than the olive. One of the oldest cultivated fruit crops, it has long been harvested for its tasty, fleshy fruit which is a staple food for many people across North Africa and Arabia. There are many hundreds of varieties of this species, each grown for commercial purposes, perhaps making the Date Palm the second most familiar palm species after the Coconut Palm *Cocos nucifera*. It grows with an imposing, tall, slender, straight trunk, which has a spiralling pattern on the bark, with long, feather-like leaves, which are greenish-grey in colour and have spines on the lower third of the stem. On the upper part of the crown, the leaves stand pointing upwards, but on the lower part, the leaves curve towards the ground. The leaves are rigid, long and pointed, with as many as 200 leaflets growing on each side of the stem. The flowers are clustered into elongated, sheathed inflorescences borne on separate male and female plants. The male's are white and fragrant, and the female's smaller, and more yellowish or cream in colour. The sugar-rich fruit, which is commonly known as the date, is a large, oblong berry that is dark orange when ripe, and may grow up to 7.5 cm (3 in) in length on some cultivated varieties.

2 Chronicles ch. 28, v. 15 records that Jericho is 'the city of palm trees'. Ezekiel chh. 40 & 41 record his vision of a new temple and how a great deal of the structure and carving was with palm wood. But there are only two references to palms in the New Testament. John the Gospel writer describes how in his later great vision he:

 REVELATION
CH. 7, V. 9

looked and there before me was a great multitude that no-one could count, from every nation, tribe, people and language, standing before the throne and in front of the Lamb. They were wearing white robes and were holding palm branches in their hands.

But to Christians throughout the world the most memorable reference to palms is at the start of the Easter story:

ST JOHN
CH. 12, VV. 12-14

The next day the great crowd that had come for the Feast [of the Passover] heard that Jesus was on his way to Jerusalem. They took palm branches and went out to meet him, shouting, "Blessed is he who comes in the name of the Lord!" "Hosanna!" "Blessed is the King of Israel!"

All four Gospel writers tell of Jesus' triumphant entry into Jerusalem on what Christians call Palm Sunday, the week before the following Sunday, which is Easter Day. But only John names palm branches.

Coconut Palm

The third tree in the list that God named was one the children of Israel found when they reached the Promised Land and found that the hills were covered with forest, much of it dense thickets of Myrtle *Myrtus communis*. This tree does not need much water, unlike the willow, and sprigs of it will remain fresh for weeks. It is an aromatic plant and may grow more than 5 m (about 16 ft) high. The leaves are

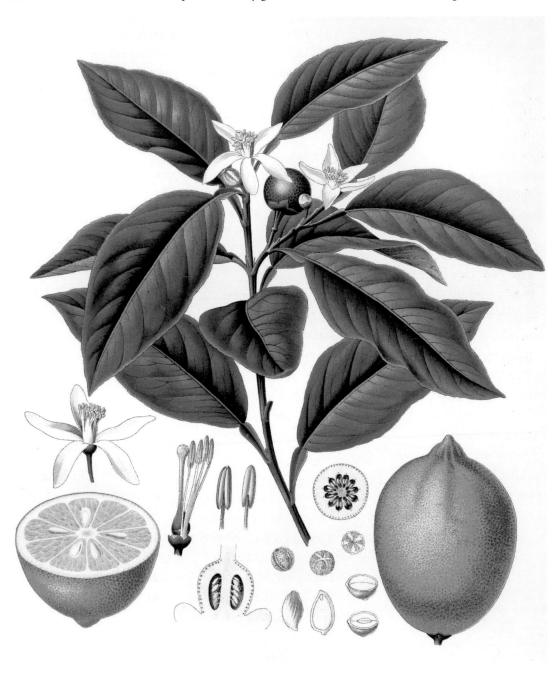

CITRON

thick and lustrous with many small, oil-bearing glands. The solitary white flowers, about 1.8 cm (about ¾ inch) long, are borne on short stalks. The fruit is a purplish black, many-seeded berry. The tree soon became a favourite. Isaiah, for example, recorded God's promise that '…. instead of briers the myrtle will grow' (Isaiah ch. 55, v. 13) and Zechariah says:

MYRTLE

ZECHARIAH
CH. 1, V. 8

During the night I had a vision – and there before me was a man riding a red horse! He was standing among the myrtle trees in a ravine.

This suggests that God's messenger (for that is who the prophet saw) was among well known and common trees in the country, trees that Zechariah knew well, and whose fragrance he would have been familiar with, as dried flowers and leaves, and perfume. Myrtol, a volatile oil found in most parts of the plant, was formerly used as an antiseptic and tonic. In ancient times the Myrtle was a symbol of immortality, and so by extension a symbol of prosperity and success. The original settlers in the Promised Land clearly felt God was giving them a reminder of this and the hope for a successful harvest next year.

Finally, the fourth tree is believed to be the Citron *Citrus medica*, a relative of the lemon, and one of the four original citrus fruits from which all types developed. Rabbinical tradition is that the Israelites brought it to Palestine when they fled from Egypt. Josephus in his *History* also believed it was the Citron. The fruit is larger than a lemon and is used as a conserve, and as an essential oil in perfumery.

In many cultures today people plant a tree as a memorial to a significant event in their lives. So it was in Abraham's life. He moved into the region of the Negev and met Abimelech, King of Gerar, on the edge of Philistine territory. The king tried to take Sarah for his wife (Abraham had said she was his sister not his wife; the story is worth reading!), but the confusion was resolved and a treaty was sworn between them.

GENESIS
CH. 21, VV. 31-33

So that place was called Beersheba, because the two men swore an oath there.
After the treaty had been made at Beersheba, Abimelech and Phicol the commander of his forces returned to the land of the Philistines. Abraham planted a tamarisk tree in Beersheba, and there he called upon the name of the Lord, the Eternal God.

TAMARISK

The Tamarisks *Tamarix* spp. are small shrubs or trees which can thrive in arid regions, especially in saline soils. Their rather feathery, grey-green foliage on leafy branches provides welcome shade, and dense spikes of pink flowers make it a noticeable feature of the landscape. They often form dense thickets.

Saul had an angry meeting with his commanders under a Tamarisk tree when he learned that David, who had fled Saul's anger, had been discovered (Genesis ch. 22), and after Saul's death in battle, he was buried under a Tamarisk at Jabesh, south of the sea of Galilee (1 Samuel ch. 31).

Today when someone receives an unexpected gift that fulfils a need – such as some money or a prize – the cry goes up, "Manna from heaven!" Many people will understand what is meant, but may not really know what 'manna' is. The answer is about 3,500 years old. The Israelites had been captives in Egypt for many years and had at last have been freed by Pharaoh, had miraculously crossed the Red Sea, which then drowned the pursuing Egyptian charioteers, and been led by Moses into the desert, where they then suffered from a lack of food.

In the desert the whole community grumbled against Moses and Aaron. The Israelites said to them, "If only we had died by the Lord's hand in Egypt. There we sat around pots of meat and ate all the food we wanted, but you have brought us out into this desert to starve this entire assembly to death."
Then the Lord said to Moses, "I will rain down bread from heaven for you. The people are to go out each day and gather enough for that day."

EXODUS
CH. 16, VV. 2-4

Moses and Aaron told the people this good news and the later promise of God that "at twilight you will eat meat, and in the morning you will be filled with bread."

That evening Quail came and covered the camp, and in the morning there was a layer of dew around the camp. When the dew was gone, thin flakes like frost on the ground appeared on the desert floor. When the Israelites saw it they said to each other, "What is it?" For they did not know what it was.
Moses said to them, "It is the bread the Lord has given you to eat."

EXODUS
CH. 16, VV. 13-15

Almond Flowers and Nut

Clearly this was not bread as we know it, but it is one of the first times we read in the Bible about a named food. As a miraculous example of God's bounty, it is a fitting start to our thinking about what the Israelites ate in the way of fruits, cereals and vegetables. Our word 'manna' comes from the Hebrew words that the Israelites spoke when they grumbled to Moses: 'man hu' which simply means 'what is it?'

Our understanding of what it might be has changed over the years. A traditional explanation was that it was a sweet secretion from the Tamarisk Tree. More recent study has discovered that this 'manna' is produced by the scale insects that feed on Tamarisks. The insects consume huge amounts of sap to obtain carbohydrates, and excrete the excess as 'honeydew'. The white, frosty looking residue was:

NUMBERS
CH. 11, VV. 7-8

like coriander seed, white, and the taste of it was like wafers made with honey, and the people ground it in mills or beat it in mortars, then boiled it in pots and made cakes of it; and the taste of it was like the taste of cakes baked with oil.

A huge amount of this would have been needed to feed the Israelites, which to many readers has seemed ridiculous. We must remember, however, that a perfectly natural explanation misses the point that both the manna and the quails in the desert were a miraculous gift from God. That wonderful providence is recalled by Jesus when he called Himself the bread of life (St John ch. 6, vv. 49 –51).

Long before the Exodus from Egypt and the miraculous manna, is the story of the Jews' arrival there. There was famine in Palestine and Joseph's brothers are sent to Egypt by their father, Jacob, to buy grain. They meet Joseph but do not recognise him. He says he will help if the youngest brother is brought to him. So a second journey is made. Jacob says:

GENESIS
CH. 43, V. 11

If it must be then do this: Put some of the best products of the land in your bags and take them down to the man as a gift – a little balm, and a little honey, some spices and myrrh, some pistachio nuts and almonds.

Almonds are often mentioned in the Bible. The Almond *Prunus dulcis* grows up to 10 m (33 ft) tall and bears fruit in the third year. It is native to Palestine and the Middle East and is often the first tree to put forth blossom, before the leaves appear, as early as January. Carvings of almond flowers were on the woodwork of the tabernacle, which held the Ark of the Covenant. The beauty of the pale pink Almond is probably the species mentioned in *Song of Songs* and in *Jeremiah*. The Almond tree is revealed to be even more important in the story of the Israelites' anger with Moses and Aaron and the revolution stirred up by Korah of the tribe of Levi against Moses and Aaron. Moses says God will sort the problem out, and so:

Pistachio

> *The Lord said to Moses, "Speak to the Israelites and get twelve staffs from them, one for each of the ancestral tribes. Write the name of each man on his staff. On the staff of Levi write Aaron's name, for there must be one staff for the head of each ancestral tribe. Place them in the Tent of the Meeting in front of the Testimony, where I meet with you. The staff belonging to the man I choose will sprout, and I will rid myself of this constant grumbling against you by the Israelites. … The next day Moses entered the Tent of the Testimony and saw that Aaron's staff, which represented the house of levi, had not only sprouted but had budded, blossomed and produced almonds.*

NUMBERS
CH. 17, VV. 1-5 & 8

So there was now no uncertainty. The importance of the role of Aaron the priest and his sons in the worship of Israel was firmly grounded.

Clearly it is a fruit tree that has played a sacred part in Jewish life. It is thought that the tree was first cultivated in ancient times in the Levant, so Jacob would have been familiar with groves of the trees. On the other hand, the command of Jacob to pack Pistachio nuts *Pistacia vera* is the only reference to this tree in the Bible. It is native to the Middle East, including Israel, and has been cultivated for many centuries. At an archaeological dig in the Hula Valley, in Israel, seeds and nut-cracking tools were found and dated to 78,000 years ago!

Soon after Almond trees blossom, the scarlet blooms of the Pomegranate *Punica granatum* come out. It was cultivated since ancient times from Egypt to Mesopotamia. When Moses talked to God and received the Ten Commandments, he was given a detailed description of the design of the priest's robes that Aaron had to wear when he entered the Holy Place:

EXODUS
CH. 28, VV. 33-35

> *Make pomegranates of blue, purple and scarlet yarn round the hem of the robe, with gold bells between them. The gold bells and the pomegranates are to alternate round the hem of the robe. Aaron must wear it when he ministers.*

The ripe fruit contains hundreds of tasty seeds. In the Middle East it had long been considered to be a fertility symbol, and its sacred significance became part of the Israelites' beliefs when they made Aaron's robes.

Later, scouts sent out by Moses to see what the 'Promised land' looked like, came back with various fruits:

POMEGRANATE

they cut off a branch bearing a single cluster of grapes.
Two of them carried it on a pole between them, along with
some pomegranates and figs.

Clearly the size of the grape cluster and the variety of fruits were indications of the goodness of the land into which God was leading them. Furthermore, pomegranates were listed by God when he said to Moses:

Olive Flowers and Fruit

Observe the commands of the Lord your God, walking in

≈ DEUTERONOMY   *His ways, and revering Him. For the Lord your God is*
CH. 8, VV. 6-9   *bringing you into a good land – a land with streams and*
pools of water, … a land with wheat and barley, vines and fig
trees, pomegranates, olive oil and honey; a land where bread
will not be scarce and you will lack nothing. ≈

This list is known as the Seven Species, being the special products of the land of Israel. Pictures of the fruit appeared on ancient Judaean coins, and decorative silver globes shaped like pomegranates sometimes cover the handles of scrolls of the Torah.

The juice of the fruit became a favourite soft drink in the summer months, as is made clear by the Beloved to her Lover:

I would lead you

≈ SONG OF SONGS   *and bring you to my mother's house –*
CH. 8, V. 2   *she who has taught me.*
I would give you spiced wine to drink,
the nectar of my pomegranates ≈

The olive oil mentioned in the quotation from Deuteronomy above leads us to consider another of the 'seven', the Olive *Olea europaea*. The wild olive grows in the groves of Upper Galilee and Carmel. It is a prickly shrub producing small fruits. There are many varieties of cultivated olives, some being suitable for oil, and some for food as preserved olives. Its foliage is dense and when it becomes old, the fairly tall trunk acquires a unique pattern on its bark. There are trees in Israel estimated to be 1,000 years old, such as the one in the garden of Gethsemane in Jerusalem. In old age the tree becomes hollow but the trunk continues to grow thicker, at times achieving a circumference of 6 m (20 ft). The olive tree blossoms at the beginning of summer and its fruit ripens about the time of the early rains in October. The fruit, which is rich in oil, is first green, but later becomes black.

Olive trees have always been the most extensively distributed and the most conspicuous in the landscape of the Mediterranean region. Production of olive oil has been traced as far back as 2400 BC in clay documents found near Aleppo in Syria. Today Israel produces only about half of the oil it needs each year, so imports are vital.

King Sennacherib, who besieged Jerusalem in 701 BC, also made use of a description like the one in Deuteronomy when promising the inhabitants of Jerusalem that he would exile them to a country of like fertility:

Make peace with me and come out with me. Then every
one of you will eat from his own vine and fig tree and
2 KINGS
drink water from his own cistern, until I come and take
CH. 18, V. 32
you to a land like your own, a land of grain and new wine,
a land of bread and vineyards, a land of olive trees and
honey. Choose life not death!

The Israelites did not leave, but Isaiah the prophet told their king, Hezekiah, that eventually, because of their sins, they would be exiled to Babylon – and that did happen, in 586 BC.

The bounty of Israel is frequently described as 'corn, wine and oil' (Deuteronomy ch. 7, v. 13, et al.), that is, grain, vines and olives, which formed the basis of Israel's economy. When the Israelites conquered the land they found extensive olive plantations (Deuteronomy ch. 6, v. 11). Western Galilee, the territory of Asher (who was Jacob's son), was – and is – especially rich in olives, so much so that Asher will 'bathe his feet in oil' (ch. 33, v. 24). They flourish in mountainous areas too, even among the rocks, thus producing 'oil out of the flinty rock' (Deuteronomy ch. 32, v. 13). Outside the walls of Jerusalem is 'The Mount of Olives' (Zechariah ch. 14, v. 4). It is named in Hebrew, Har ha-Mishhah, 'The Mount of Oil'. The fruit of the olive also develops well in the plain between Mt Hermon and the coast (the Shephelah Lowland), where it grows near sycamores. These crops were so precious that David appointed special overseers to manage the plantations. So far as we know David did not raise taxes; he financed his court by wealth from his extensive land holdings, commerce, plunder and tribute from subjugated kingdoms – so he depended on his orchards being well looked after:

1 CHRONICLES
Baal-Hanan the Gederite was in charge of the olive and
CH. 27, V. 28
sycamore trees in the western foothills. Joash was in
charge of the supplies of olive oil.

The olive tree is full of beauty, especially when laden with fruit, as the prophet tells us: 'a leafy olive-tree, fair with goodly fruit' (Jeremiah 1, v.16). It is an evergreen, and the righteous who take refuge in the protection of God are compared to it:

PSALM
But I am like an olive tree
52, V. 10
flourishing in the house of God.

As we have read, the Olive is long lived, sometimes growing into a marvellous, twisted, gnarled tree. The 'olive shoots' referred to in Psalm 128 are actually the saplings that sprout from its roots and protect the trunk and, if it is cut down, they ensure its continued existence:

Your wife will be like a fruitful vine within your house; your sons will be like olive saplings round about your table.

The wood is very hard and beautifully grained, making it suitable for the manufacture of small articles and ornaments; the hollow trunk of the adult tree, however, renders it unsuitable for pieces of furniture. Some commentators, therefore, say that the olive cannot be the wood (*ez shemen*) from which the doors of Solomon's Temple were made:

1 KINGS
CH. 6, V. 31

For the entrance of the inner sanctuary he made doors of olive wood with five-sided jambs.

The entrance to the main hall had similar doors from olive wood, so the scripture says.

In spring the olive tree is covered with thousands of small whitish flowers, most of which fall off before the fruit forms. Job's 'comforter', Eliphaz the Temanite, uses this image of the tree in spring, to tell Job that if he is not wise:

JOB
CH. 15, V. 33

He will be like a vine stripped of its unripe grapes, like an olive tree shedding its blossoms.

The great value of olives is described by Zechariah in a vision where he tells that there are two olive trees in the Temple which stand for the priestly and royal offices, and symbolize a constant supply of oil (Zechariah ch. 4). After ripening, the fruit is harvested in two different ways, by beating the branches with sticks or by hand picking. The former way is quicker but many branches are broken and this diminishes successive harvests. This method was used in biblical times, the Bible commanding that the fruits on the fallen branches are to be a gift to the poor:

When you beat the olives from your trees, do not go over the branches a second time. Leave what remains for the alien [immigrants], *the fatherless and the widow.*

The same law also told the farmer not to go over the vines again, for the same reason (Deuteronomy ch. 24, vv. 20, 21). The second method was the more usual from about 200 AD onwards according to the early part of the Talmud. It was termed masik ('harvesting olives'), the fingers being drawn down the branches in a milking motion so that the olives fall into the hand. By this method the harvested olives remained whole, whereas by the other the olives were often bruised by the beating. There were olives of different varieties and different sizes used, for oil in cooking and in lamps, in preserving, and in religious rituals. Despite these differences, the olive was designated as a standard size for many laws, and the expression 'land of olive trees' was interpreted as 'a land whose main standard of measurement is the olive'. Today we can buy olive oil or virgin olive oil. The latter will have been produced by traditional, mechanical crushing methods, but non-virgin oils will very likely have been chemically treated, to improve the taste, for example.

For centuries the olive has been used in cooking, and in religious practices for healing, strength and consecration. Perhaps the best known example in the Old Testament of the last is when Samuel, after the death of King Saul, is led by God to anoint the youngest son of Jesse:

 1 SAMUEL
CH. 16, V. 16

So Samuel took the horn of oil and anointed him in the presence of his brothers, and from that day on the Spirit of the Lord came upon David in power.

In the New Testament, the story of an unnamed woman at the home of Simon the Pharisee is particularly well known in the Christian faith. Jesus says to Simon:

ST LUKE
CH. 7, VV. 44-46

I came into your house. You did not give me any water for my feet [which in these times was the minimal gesture of hospitality to show a visitor who had walked a dusty road in sandals]. *You did not put oil on my head, but she has poured perfume on my feet.*

The oil's everyday use in the kitchen is told in the story from the ninth century BC, of Elisha the prophet and the widow's jar of oil, which also tells of the importance and power of olive oil. Her husband has died and his creditor is coming to take her two boys away. She cries out to Elisha for help, and he tells her to collect as many jars as she can and pour her little bit of oil into all the jars she collects. She fills every one!

2 KINGS
CH. 4, V. 7

She went and told the man of God and he said, "Go, sell the oil and pay your debts. You and your sons can live on what is left."

Finally, the modern symbol of peace, an olive leaf in the beak of a dove, can trace its origin back to the story of Noah's dove which returned to the Ark carrying an olive leaf (Genesis ch. 8, v. 11).

Figs are recorded in both the Old and New Testaments. To cover their nakedness and shame in the Garden of Eden, Adam and Eve sewed fig leaves together; King Sennacherib promised each Israelite could have his own fig tree; King David appointed Baal-Hanan the Gederite as overseer of his

Fig Leaves and Fruit

sycamores; Jesus told a parable about a fig tree, and Zacchaeus climbed one. They were all referring to fruit from the same family but from two different trees: the Common Fig *Ficus carica* and the Sycomore-fig *F. sycomorus*. The Fig Tree is the first to be named with a common name in the Bible, and the third of all trees to be mentioned after the Tree of Life and the Tree of the Knowledge of Good and Evil (Genesis ch. 3, v. 7).

Common Figs were (and are) cultivated throughout Palestine. A summer tree in full leaf gives cool, welcome shade, which is almost certainly why Micah wrote in about 700 BC that:

<table>
<tr><td>MICAH
CH. 4, V. 4</td><td>*Every man will sit under his own vine*
and under his own fig tree,
and no-one will make them afraid,
for the Lord Almighty has spoken.</td></tr>
</table>

Failure of their harvest would cause great distress, whether by natural or warlike means:

<table>
<tr><td>JEREMIAH
CH. 5, VV. 15-17</td><td>*"O house of Israel," declares the Lord,*
"I am bringing a distant nation against you –
an ancient and enduring nation...
they will devour your flocks and herds,
devour your vines and fig-trees."</td></tr>
</table>

The trees lose their leaves in winter and by the end of March are in bud again. Tiny figs form at the same time as the leaves appear, grow to about the size of a cherry, then the majority of them are blown by the wind to the ground. These 'green', 'untimely' or 'winter' figs are collected and can be eaten. Some do develop and become 'very good figs, like those that ripen early' (Jeremiah ch. 24, v. 2). They are known by Micah as 'the early figs that I crave' (Micah ch. 7, v. 1). As these are ripening, little buds of the next crop form further up the branches, to be harvested in August.

Besides being eaten fresh or dried they are also pressed into a solid cake (1 Chronicles ch. 12, v. 40) which can be cut with a knife, as was prepared by Abigail for David before he became king (1 Samuel ch. 25, v. 18); or even used medicinally, as was prescribed for King Hezekiah (2 Kings ch. 20, v. 1).

The other fig, the Sycomore-fig, is recorded several times, firstly when David appointed Baal-Hanaan as warden of his orchards, and later when Solomon made 'cedar as plentiful as Sycomore-fig trees in the foothills' (1 Kings ch. 10, v. 27). These are large trees up to 20 m (65 ft) tall, with a dense crown of spreading branches. They are mostly found in Africa south of the Sahara, but long ago were naturalized in Egypt and Israel and grown as an orchard fruit tree. Evidence of the cultivation of Sycomore-figs in Old Testament times is told by the prophet Amos when he describes his occupation to Amaziah the priest:

I was neither a prophet nor a prophet's son, but I was a shepherd, and I also took care of Sycamore-fig trees. ✑

To get the tree to bear good, ripe fruit is difficult. It depends on a complex association with insects and other creatures, and the gardener having to split the top of each fig. The obscure Hebrew word to describe Amos's work is variously translated as 'dresser of', 'tend' or 'took care of', and physical evidence of the work has been found in the tombs of Ancient Egypt where the fruit was cultivated extensively from the start of the third millennium BC.

It is perhaps the best-known tree in the New Testament due to the story found only in St Luke's gospel, its wide canopy certainly being sturdy enough for a grown man to climb high into it:

✑ ST LUKE
CH. 19, VV. 1-4

Jesus entered Jericho and was passing through. A man was there by the name of Zacchaeus; he was a chief tax collector and was wealthy. He wanted to see who Jesus was, but being a short man he could not, because of the crowd. So he ran ahead and climbed a Sycomore-fig Tree to see him, since Jesus was coming that way. ✑

He did see Jesus, but more to the point Jesus saw him. Although Luke writes that Jesus was originally just passing through Jericho, He makes a point of speaking to Zacchaeus, and inviting Himself to stay at his house! Zacchaeus was thrilled, although the crowd was not: "He's gone to be the guest of a 'sinner' ". Such men in the employ of the Romans for extorting money from the Israelites, were all condemned as far as the faithful were concerned. But Jesus recognized a wonderful opportunity to preach one of His most important messages. Zacchaeus repented of his greed and promised to repay what he had gained by cheating:

✑ ST LUKE
CH. 19, VV. 9-10

Jesus said to him, "Today salvation has come to this house, because this man, too, is a son of Abraham. For the Son of man came to seek and to save what was lost." ✑

Jesus saw that Zacchaeus was a true Jew, even though the crowd did not. Jesus' last sentence quoted above is a splendid summary of His purpose here on earth – to bring salvation to all, which meant eternal life and being part of the kingdom of God.

It is worth pointing out that the tree was a Sycomore-fig, not a Sycamore (note the spelling) which many people in Britain know as a wild tree, and was the word used in the Authorized Version of the Bible. Modern scholarship has corrected later translations.

In another story, the Parable of the Prodigal son, Jesus mentions yet another fruit. The son had

Carob Tree and 'Locust Beans'

selfishly taken his inheritance and left home for a foreign land and squandered all his money.

ST LUKE
CH. 15, VV. 14-16

After he had spent everything there was a severe famine in the whole country, and he began to be in need. So he went and hired himself out to a citizen of that country, who sent him to his fields to feed pigs. He longed to fill his stomach with the pods that the pigs were eating, but no-one gave him anything.

Once again Jesus preaches a powerful message about the fact that God loves a sinner who repents. The point of His message is made especially sharp by having the boy live with pigs, which to Jews are unclean, so to the Jews listening to Jesus, the boy was a particularly bad sinner.

The pods were almost certainly the fruits of the Carob Tree *Ceratonia siliqua*. It is also known as The Locust Bean and St John's Bread. These traditional names may mean that John the Baptist did not eat honey and locusts in the wilderness, but ate honey and locust-beans. It is a native of the Middle East and has been cultivated since ancient times. The use of the Carob during a famine is likely a result of the tree's resilience to the harsh climate and drought. During a famine, the swine were given Carob pods so that they would not be a burden on the farmer's limited resources. As many will well know there is a happy ending. The son is welcomed home by his father, and is forgiven. Today the pod is still important commercially. It is used as a thickening agent in the food industry, and in the production of biscuits and cakes, and as a chocolate substitute.

From the first book in the Bible to the last, the Grape *Vitis vinifera* is mentioned. Noah, a farmer, was the first person recorded to plant a vineyard after he, his family and the animals came on land again after the flood (Genesis ch. 9, vv. 20–23). He also became the first person to drink too much and become drunk, much to the distress of his sons, Ham, Shem and Japheth. Noah prayed that God would bless his sons and make them prosper. A later writer emphasised wine's power when he wrote 'Don't let wine tempt you, even though it is rich red (Proverbs ch. 23, v. 21). Another Old Testament book has an erotic image spoken by a male lover to his beloved, which names our fruit and its tempting power:

SONG OF SONGS
CH. 7, VV. 6-9

How beautiful you are and how pleasing,
O love, with your delights!
Your stature is like that of the palm,
and your breasts like clusters of fruit.
I said, "I will climb the palm tree;
I will take hold of its fruit."
May your breasts be like clusters of the vine,

A Cluster of Grapes

the fragrance of your breath like apples,
and your mouth like the best wine. ≈

St John, writing in Revelation, the last book, sees a vision of the final judgement of the people of the world. One angel gathers all the righteous people, and another is told to harvest the rest:

≈ REVELATION
CH. 14, vv. 18-20

"Take your sharp sickle and gather the clusters of grapes
from the earth's vine, because the grapes are ripe."
The angel swung his sickle on the earth, gathered its grapes
and threw them into the great winepress of God's wrath.
They were trampled in the winepress outside the city,
and blood flowed out of the press ..." ≈

Whatever may be our understanding of this apocalyptic description of the end of the world, it does tell us how the farmers of that time extracted the juice of the grapes. The winepress was a rock-hewn trough about 2.5 m (8 ft) square with a channel on one side leading to a smaller trough. Grapes were thrown in the upper trough and trampled with bare feet, and the juice flowed into the lower trough, to be collected and turned into wine.

The Old Testament has dozens of references to grapes and vineyards and wine, which shows how important it was at that time. But wine was not just for a drink, it was to be part of the regular sacrificial offerings as recorded in Exodus ch. 29, Leviticus chh. 1–7 and Numbers ch. 28. But above all it is the Psalmist who sums up the Jews' feelings about the vine, the grape and the wine that gladdens the heart of man (Psalm 104, v. 14).

To the Jews the vineyard and what grew there were more than just a source of sweet food to gladden his heart. The prophet Isaiah (749–681 BC) put it like this in a section of his prophecy often titled as The Song of the Vineyard:

≈ ISAIAH
CH. 5, vv. 1-2 & 7

I will sing for the one I love
a song about his vineyard:
My loved one had a vineyard
on a fertile hillside.
He dug it up and cleared it of stones
and planted it with the choicest vines.
He built a watchtower in it
and cut out a winepress as well.
Then he looked for a crop of good grapes
but it yielded only bad fruit...

The vineyard of the Lord Almighty
is the house of Israel;
and the men of Judah
are the garden of his delight.
And he looked for justice, but saw bloodshed;
for righteousness, but heard cries of distress. ❧

Isaiah describes a special vineyard and then interprets it with a powerful play on words in the last two lines – in Hebrew the words for 'justice' and 'bloodshed' sound alike, as do 'righteousness' and 'distress'. There follows Isaiah's declaration of God's judgement on the sinful nation of Judah which will come, including the fact that 'a ten-acre vineyard will produce only a bath of wine' (i.e. about 36 litres (8 gallons), Isaiah ch. 5, v. 10), which is a very small fraction of what the harvest should be. The total depends on the quality of the grapes, the quality of the land, the density of the planting and careful husbandry; then good wine can be produced from 4–8 tons of grapes per acre, and that can give about 700 litres (150–160 gallons) of wine per ton per acre. God's punishment is a really harsh one. Jews today may believe that God's vineyard of Israel is the conquering one now, not the conquered nation Isaiah was describing, but it is certainly not peaceful.

Jesus was also adept at creating telling images from everyday life. One of His most memorable is His Parable of the Tenants of a Vineyard, which is found in Matthew, Mark and Luke, is probably based on *The Song of the Vineyard*. St Matthew tells the parable in detail. The tenants of the vineyard were greedy and thought they could make it their own. They killed a servant, beat and stoned others, and then the owner sent his son, who surely would be respected. But he, too, was killed. This parable was told by Jesus in Jerusalem a little while after He had angrily overturned the tables of the money-changers in the temple. It is one of His statements at this time warning the disciples of His imminent death. For Christians, this parable is a constant reminder that Jesus Christ does indeed hold the church together. John sums up much of what Christians believe, when he records these words of Jesus:

❧ ST JOHN
CH. 15, VV. 1-3 & 16

I am the true vine, and my Father is the gardener.
He cuts off every branch in me that bears no fruit, while
every branch that does bear fruit he prunes so that it will be
even more fruitful. You are already clean [i.e. pruned]
because of the word I have spoken to you. Remain in me, and
I will remain in you. No branch can bear fruit by itself; it
must remain in the vine. Neither can you bear fruit unless
you remain in me… I chose you and appointed you to go
and bear fruit – fruit that will last. ❧

The disciples and people today still puzzle over what is meant by 'fruit'. Some think it just means that you have persuaded someone to be a Christian. I think it is to be more widely understood as described by St Paul:

 GALATIANS
CH. 5, VV. 16, 22, 23

So I say, live by the Spirit, and you will not gratify the desires of sinful nature... But the fruit of the Spirit is love, joy, peace, patience, kindness, goodness, faithfulness, gentleness and self-control. Against such things there is no law.

That is a rich harvest of fruit indeed, nine grapes in one cluster, giving wine of the best quality! That is the harvest which will enrich the Christians' lives and the church's evangelism, and ensure all believers do love their neighbours as themselves.

CHAPTER III

CEREALS, HERBS
AND FLOWERS

Today agriculture in Israel is led by modern technologies, and she exports a considerable amount of fresh produce, despite the fact that about half the country is desert, and the climate means there is a shortage of water. After the story of creation, the Bible records that the story of the Jews begins with farmers. The first word in the Hebrew text is *hereshith* ('in [the] beginning'). This is the Hebrew title of the book (our name comes from the Greek Septuagint translation), so the early history of the Hebrew world is there. After the stories of the creation and the Garden of Eden we come to a family tale:

GENESIS
CH. 4, VV. 1 & 2

Adam lay with his wife Eve, and she became pregnant and gave birth to Cain. She said, "With the help of the Lord I have brought forth a man." Later she gave birth to his brother Abel. Now Abel kept flocks, and Cain worked the soil.

Eventually they each bring an offering to the Lord of what they have produced. Cain brought 'some fruits' but Abel 'brought fat portions from some of the first born of his flock'. God looked favourably on Abel's generous offering but He did not look favourably on Cain's casual gift. Cain was jealous of his brother's being praised by God, killed him, and when God asked where his brother was, Cain's reply has entered the English language as a saying from someone who claims to have nothing to do with the welfare of a sibling or acquaintance: "Am I my brother's keeper?" (v. 9). Sadly, this callous indifference to others is still widespread in the world today, and results in individuals and nations being like Cain – outcasts.

Archaeology has revealed that the early history of the people of the region did indeed keep flocks or grow crops. The herders kept sheep, goats, asses and 'cattle'. In The Old Testament the last name is often a collective noun for all the types of animal. Abraham, Isaac and Jacob were such people, pastoral farmers, semi-nomadic, wandering to find food for their animals. On the other hand, the people of Egypt, thanks to the fertile soil along the Nile valley, were famous for their crops of grain. By Christ's time when the Romans ruled the world, Egypt was 'the bread basket' of the known world. Over the course of centuries growing crops became increasingly important, and constantly in the Old Testament the listing together of grain, wine, oil and pulses shows that these products were the principal harvests of the Palestinian farmer, and the Jewish people's social and religious year revolved around the several harvest periods.

Wheat, Barley and Oats

In 1908 archaeologists digging at the ancient city of Gezer west of Jerusalem found an inscribed limestone tablet, which has become known as the Gezer Calendar:

two months gathering [olives] (September, October)
two months planting (November, December)
two months late sowing (January, February)
one month cutting flax (March)
one month reaping barley (April)
one month reaping and measuring grain (May)
two months pruning [vines] (June, July)
one month summer fruit (August)

Whether these are the words of a popular song or a schoolboy's memory exercise does not hide the fact that clearly the Palestinian farmer was toiling throughout the year. Besides the tasks listed he also had to clear the stony land ready to plough, and then had to irrigate the land during the dry season, as is recorded in *The Apochrypha* by bringing

◁ ECCLESIASTICUS
CH. 24, vv. 30-31

a brook from a river and ... a water channel into a garden ◁

The descendants of Cain grew barley and wheat. The first reference to barley is in the recording of the dreadful catastrophes that overtook Egypt because Pharaoh refused to let the Israelites leave the country. God told Moses to tell Pharaoh:

◁ EXODUS
CH. 9, vv. 18 & 31-32

"... at this time tomorrow I will send the worst hailstorm that has ever fallen on Egypt, from the day it was founded till now"... The flax and barley were destroyed, since the barley was in the ear and the flax was in bloom. The wheat and spelt however, were not destroyed, because they ripen later. ◁

We will consider the two cereal grains in the order they are mentioned, which is not the same as their importance or value, as we shall see. Barley *Hordeum vulgare* is a cereal grain, a self-pollinating member of the grass family. Its wild ancestor grew abundantly in the Fertile Crescent of western Asia, a horseshoe shape of land from Mesopotamia to Upper Egypt. This is the very land of Cain and the Patriarchs of old, so it is perfectly natural that we should find that one of God's laws for them says:

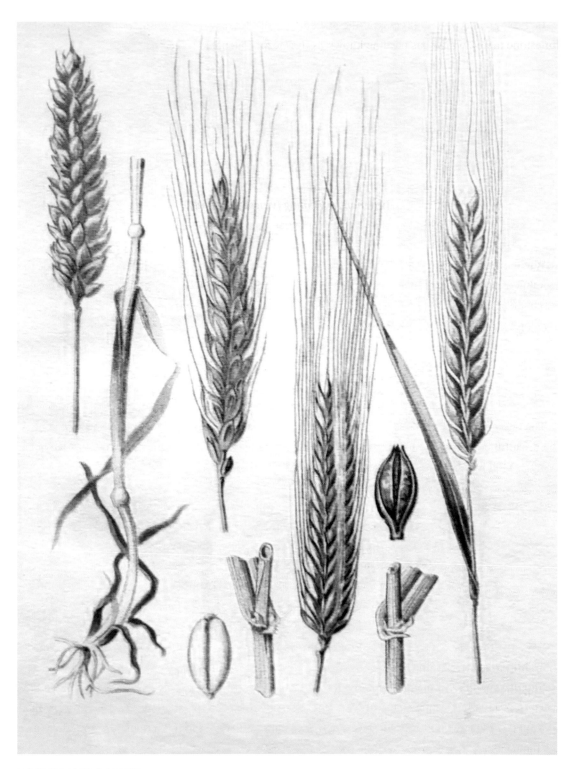

Wheat and Barley

	If a man dedicates to the Lord part of his family land,
LEVITICUS	its value is to be set according to the amount of seed
CH. 27, V. 16	required for it – fifty shekels of silver to a homer
	(6 bushels or about 220 litres) of barley seed.

In ancient times the shekel was simply a unit of weight of about 11 gm ($\frac{1}{3}$ oz), and it is thought that it originally referred to the weight of Barley. Later it became a coin's name as well, and today in Israel it refers only to that country's currency. Barley was harvested in April or May, depending on the weather and the farm's position.

The size of the barley harvest in about 950 BC can be judged from words that King Solomon sent to King Hiram of Tyre as part of his request for wood and workers to help build his new temple:

	My men shall work with yours to provide me with
	plenty of timber, because the temple I build must be large
2 CHRONICLES	and magnificent. I will give your servants, the woodsmen
CH. 2, VV. 8-10	who cut the timber, 20,000 cors [c. 4,400 kilolitres/
	120,000 bushels] of ground wheat, 20,000 cors of barley,
	20,000 baths [c. 440 kilolitres/95,000 gallons] of wine
	and 20,000 baths of olive oil.

Hiram agreed to supply the timber, appointed a skilled craftsman to work with Solomon's men, and sent the timber. The quantity of food and drink that Solomon promised was much needed – the end of chapter 2 records that he ordered 153,000 men to be the work force! By New Testament times it is recorded that a worker gets 'three quarts (about a litre) of barley for a day's wage' (Revelation ch. 6, v. 6).

Barley produces a nutritious food. Although it was considered an inferior grain to wheat, its wholemeal flour was and is much used in cooking. After the death of Joshua, Israel fell into the hands of the Midianites. Their home territory was flanking the eastern arm of the Red Sea. It was where Moses fled after he had killed the Egyptian (Exodus ch. 2). After nine years of oppression, Gideon was leader of the desire to be rid of the Midianites:

	Now the camp of Midian lay below him in the valley.
	During the night the Lord said to Gideon, "get up, go down
JUDGES	against the camp, because I am going to give it into your
CH. 7, VV. 8-14	hands. If you are afraid to attack, go down to the camp with
	your servant Purah and listen to what they are saying...
	Gideon arrived just as a man was telling his friend his dream.

"I had a dream," he was saying. "A round loaf of barley bread
came tumbling into the Midianite camp. It struck the tent
with such force that the tent overturned and collapsed."
His friend responded, "This can be nothing other than
the sword of Gideon son of Joash, the Israelite. God has
given the Midianites into his hands." ✑

Indeed He had. With only 300 men Gideon routed the enemy who fled. Revelations by dreams are often mentioned in the Old Testament. This dream's imagery is particularly apposite because the barley loaf is worth only half as much as a wheat one, and is a good symbol for Israel whose numbers were far fewer than the enemy. Gideon would have found much comfort in what he overheard.

In contrast a splendid, peaceful story which includes the barley loaf, is found in all four Gospels, although only John names what sort of bread is involved:

✑ ST JOHN
CH. 6, vv. 3-13

Jesus went up on a mountain and sat down with his disciples.
The Jewish Passover feast was near.
When Jesus looked up and saw a great crowd coming towards
him, he said to Philip, "Where shall we find bread for these
people to eat?" He asked this only to test him, for he already
had in mind what he was going to do.
Philip answered him, "Eight months' wages would not buy
enough bread for each one to have a bite!"
Another of the disciples, Andrew, Simon Peter's brother, spoke
up. "Here is a boy with five barley loaves and two small fish,
but how far will they go among so many?"
Jesus said, "Make the people sit down." There was plenty of
grass in that place, and the men sat down, about five
thousand of them. Jesus then took the loaves, gave thanks,
and distributed to those who were seated as much as they
wanted. He did the same with the fish.
When they had all had enough to eat, he said to the disciples,
"Gather the pieces that are left over. Let nothing be wasted."
So they gathered them and filled twelve baskets with
the pieces of the five barley loaves left over by those
who had eaten. ✑

BARLEY

Barley loaves were cheap, the food of the poor. The miraculous feeding of the crowd of 5,000 men – women and children were not counted – was a splendid preface to the meeting that Jesus had later with His disciples when He told them that:

ST JOHN
CH. 6, VV. 26-35

"You ate the loaves and had your fill. Do not work for food that spoils, but for food that endures to eternal life.,. which the Son of man will give you... "
"Sir," they said, "from now on give us this bread."
Then Jesus declared, "I am the bread of life. He who comes to me will never go hungry, and he who believes in me will never be thirsty."

This declaration by Jesus is the first of seven I AM descriptions by Jesus of Himself, all recorded only in John's gospel. The crowd and the disciples did not understand it. They still thought Jesus was talking about real bread. With "I am the bread" Jesus is echoing God's words telling Moses His name, "I AM WHO I AM" (Exodus ch. 3, v. 14), and in so saying Jesus is stating His own divinity and putting Himself on a collision course with the authorities. He left the crowd grumbling and arguing, "How can this man give us his flesh to eat?" The disciples said it was a hard lesson, but Simon Peter finally admitted on their behalf, "You have the words of eternal life. We believe and know that you are the Holy One of God." (St John ch. 6, vv. 68, 69). Christians today still find this is a hard lesson to understand, and they need the faith of Peter to be able to say "I believe".

One of the best known stories in the Old Testament is the story of Jacob's young son, Joseph. Thanks to Andrew Lloyd Webber's musical *Joseph and the Technicolour Dreamcoat* thousands of people, young and old, know something of the story from Genesis. The story begins when

GENESIS
CH. 37, VV. 2-8

Joseph, a young man of seventeen was tending the flocks with his brothers ... Now Israel [i.e. Jacob, his father] loved Joseph more than any of his other sons because he had been born to him in his old age; and he made a richly ornamented robe for him. When his brothers saw their father loved him more than any of them, they hated him and could not speak a kind word to him.
Joseph had a dream, and when he told it to his brothers, they hated him all the more. He said to them, "Listen to this dream I had: we were binding sheaves of corn out in the fields when suddenly my sheaf rose and stood upright, while your sheaves gathered round mine and bowed down to it."

His brothers said to him, "Do you intend to reign over us?" Their jealousy resulted in their plotting to kill him, but instead when they were all out tending their flocks they sold him to Midianite traders who were on their way to Egypt. That at last brings us conclusively to the resolution of the very important story of Joseph's dream about wheat. In Egypt Joseph interpreted dreams for servants and the Pharaoh, and was rewarded by the Pharaoh who put him in charge of all Egypt, which meant he controlled the saving of the wheat harvest throughout the country:

> GENESIS
> CH. 41, V. 49

Joseph stored up huge quantities of grain, like the sand of the sea. It was so much he stopped keeping records because it was beyond measure.

The rest of the family story and its happy ending is told in Genesis ch. 42–47.

Wild *emmer* and *einkorn* forms of Wheat *Triticum* spp. were first cultivated in the Fertile Crescent. Joseph's dream 'corn' is really Wheat; in English the word 'corn' for centuries has referred to any form of grain; nowadays the word technically should refer to Maize *Zea* spp. which is a cereal from the Americas. Archaeological evidence of *einkorn* wheat has been discovered in Jordan dating back to 7500–7300 BC, and there is evidence for *emmer* wheat in Iran as old as 9600 BC. The genetics of Wheat are complicated. It is self-pollinating and cultivation has resulted in the creation of several domestic forms. In Joseph's time Wheat was harvested in early summer, about a month after Barley. Harvesting was labour intensive. The farmer would grasp a handful of stalks in his left hand, cut them fairly high up with a sickle held in the right hand, and bind them into sheaves. Corn was left growing in difficult-to-reap corners of the field; that and seed fallen to the ground was left for the gleaners, as is told in the story of Ruth. She was a Moabite woman over 3,000 years ago who asked her widowed Israelite mother-in-law if she could go in the fields and pick up the leftover grain. The rest of the story is an intimate, delightful glimpse into family life (Ruth chh. 1–4).

The farmer was left with much still to do. Sheaves were taken to a threshing floor in carts or on the backs of asses. This floor was a circular patch of hard, dry, flat ground, which was usually the common property of the village. A common method of threshing was to scatter the sheaves and drive over them with a hard wooden sledge pulled by an ox or two. The grain was now freed from the husk but was mixed with broken straw and chaff. This was thrown into the air from a shovel – this is called winnowing – whereupon the heavier grain fell and the chaff blew away. Very often a final cleaning was achieved by sieving the grain. Several times the prophet Isaiah uses images of Wheat farming to illustrate what is happening to the Jews, especially when he tells them this message from God:

WATER MELON (CENTRE)

ISAIAH
CH. 28, VV. 23-29

Listen and hear my voice;
pay attention and hear what I say.
When a farmer ploughs for planting,
does he plough continually?
Does he keep on breaking up and
harrowing the soil?
When he has levelled the surface.
does he not sow carraway and scatter cummin?
Does he not sow wheat in its place,
barley in its plot, and spelt in its fields?
His God instructs him
and teaches him the right way.
Caraway is not threshed with a sledge,
nor is a cartwheel rolled over cummin;
caraway is beaten with a rod, and cummin with a stick.
Grain must be ground to make bread;
so one does not go on threshing it for ever.

Spelt is yet another form of ancient Wheat. There is scientific evidence that it was a hybrid of *emmer* and a wild grass which was formed in the Middle East long before Wheat *Triticum aestivum* as we know it appeared. Despite the Bible reference, some commentators do not believe spelt was cultivated in Mesopotamia, but is the result of the confusion with *emmer* wheat.

Isaiah's poetic parable is thought to be saying that although God must punish Israel, his actions will be as well controlled as a good farmer's would be.

Finally the precious crop was stored. If done carefully it would keep for several years, as Joseph found in Egypt. Underground, bottle-shaped silos were dug, or the grain was kept in big earthenware jars. Modern excavations at Jericho found millet, barley and lentils in round clay bins which had been stored over 5,000 years ago! The joy of the farmer is expressed by the excited words of the Psalmist in a prayer to God:

PSALM
65, VV. 9-13

You care for the land and water it;
you enrich it abundantly.
The streams of God are filled with water
to provide the people with corn,
for so you have ordained it.
You drench its furrows
and level its ridges;

you soften it with showers and bless its crops.
You crown the year with your bounty,
and your carts overflow with abundance.
The grasslands of the desert overflow;
the hills are clothed with gladness.
The meadows are covered with flocks
and the valleys are mantled with corn;
they shout for joy and sing.

Today the harvest is still a time celebrated by people of many faiths. Christians in country and city churches sing harvest songs such as *We plough the fields and scatter/The good seed on the land*. City churchgoers sing it even though no-one in the city has ever ploughed a field or harvested the ripe corn. The believers are mindful of the debt they owe to God for the bountiful food that *is* provided. It may well have come from the supermarket, but someone at home or abroad has produced it.

Isaiah's message from God quoted above mentions other crops apart from grain. Vegetables and herbs were a part of home life but a vegetable garden as we know it was likely only at the house of the wealthy. When Ahab was king of Israel, the northern kingdom, from 874– 853, he said to Naboth in Samaria, "Let me have your vineyard to use for a vegetable garden". Naboth refused, but Ahab's wife Jezebel plotted against Naboth, had him stoned to death for reportedly having cursed God and the king, and so gained the vineyard. But Ahab would have nothing to do with his wife's evil desires (1 Kings ch. 21).

Many years later Isaiah tells the sinful Israelites that the desolation they have suffered from invaders over at least two centuries has resulted in Jerusalem being no more defensible than a

 ISAIAH
CH. 1, V. 8

shelter in a vineyard,
like a hut in a field of melons,
like a city under siege.

The owner of a vineyard often built a tower there and employed a watchkeeper to protect his crop. The word for 'melons' here is translated as 'cucumbers' in some versions of the Bible. The Watermelon *Citrullus lanatus* was once common in ancient times in Egypt; its seeds were found in Pharaoh Tutankhamen's tomb. It is now grown widely in Palestine and marketed throughout the land.

At the end of His life when Jesus came finally to Jerusalem, He taught the crowds and harangued the hypocrisy of the Saducees and Pharisees, using a homely image to make His point:

ST MATTHEW

CH. 23, V. 23

(ALSO ST LUKE CH. 11, V. 42)

Woe to you, teachers of the law and Pharisees, you hypocrites! You give a tenth of your spices – mint, dill, and cummin. But you have neglected the more important matters of the law – justice, mercy and faithfulness. You should have practised the latter, without neglecting the former.

MINT

Jesus doesn't criticise observing the law, but the hypocrisy of following some parts but not others. We sometimes do just that today, following the liturgy of our favourite church service but failing to love our neighbours as ourselves.

The stories clearly show that vegetables and spices were an important part of life then. They were all used to season food. Mint *Mentha* spp is a perennial with many forms worldwide, including Peppermint *M. piperita* and Spearmint *M. spicata* which are the two species most widely used in the temperate world in cooking. The leaf is used, fresh or dried, in Europe and the Middle East with lamb, or to make mint tea. Dill *Anethum graveolens* is related to celery. Its leaves are aromatic and are widely used in Europe, the Middle East and Asia to flavour fish, pickles and soup. It is an annual herb. Cumin *Cuminum cyminum* is an annual plant in the parsley family, and the seasoning is obtained by using the dried or crushed seeds, giving an earthy flavour. It was a feature of Ancient Egyptian cooking and later part of Jewish, Greek and Roman cuisine. Shortly before King David went to battle with Absolom who had conspired against him, the people he was with

LENTILS

2 SAMUEL
CH. 17, vv. 28-29

brought wheat and barley, flour and roasted grain, beans and lentils, honey and curds, and cheese from cows' milk for David and his people to eat.

It is suggested in *The Jewish Virtual Library* that the beans were Broad Beans *Vicia faba;* historians believe they were one of the earliest plants to be cultivated, and became part of mankind's diet in the eastern Mediterranean around 6000 BC. The Lentils *Lens culinaris* are edible seeds in the pea family, known as pulses. They have been eaten by humans since Neolithic times. Lentils were also one of the first crops cultivated in the Middle East as long as 13,000 years ago. Modern science has discovered they are a rich source of protein, and were undoubtedly a common food of the poorer people, judging from the few references we have in scripture. The most famous Bible story which speaks of Lentils is

GENESIS
CH. 25, vv. 29-34

when Jacob was cooking some stew, Esau came in from the open country, famished. He said to Jacob [his brother], "Quick, let me have some of that red stew! I'm famished!"
(That is why he was also called Edom).
Jacob replied, "First sell me your birthright."
"Look I am about to die" Esau said. "What good is the birthright to me?"
But Jacob said, "Swear to me first." So he swore an oath to him, selling his birthright to Jacob.
Then Jacob gave Esau some bread and some lentil stew.

Peppermint

When the lentil pods are boiled they turn reddish-brown, still remembered by many today in the words of the Authorized Version as 'a mess of pottage'. Later, encouraged by his mother who treated Jacob as her favourite son, he tricked his old, blind father into giving the eldest son's blessing to himself, causing much family friction. The story is a timely reminder even today of the dangers of parental favouritism.

One of the most exotic references to plants is in *Song of Songs*. The Lover says to his beloved:

HENNA

SONG OF SONGS
CH. 4, vv. 13-14

*Your plants are an orchard of pomegranates
and choice fruits, with henna and nard,
nard and saffron,
calamus and cinnamon,
with every kind of incense tree,
with myrrh and aloes
and all the finest spices.*

The substances named are a mixture of fruit which we have already discussed, cosmetics and incense. Henna *Lawsonia inermis*, or the Mignonette tree, 1.8–7.6 m (6–25 ft) tall, is found across Africa to Australia. The Lover mentions Henna early on in his praise of his beloved because she was certainly using Henna cosmetically. For over 6,000 years the paste made from dried, crushed leaves mixed with one of several liquids has long been used as a hair dye and as a paint to make intricate decorations on the skin. Mummies with red hair have been found in ancient Egyptian tombs. At Festivals such as Purim and Passover use of Henna was part of the Jewish celebrations. Some Jewish women have a Henna party a week before a wedding when the bride's grandmother paints the palms of the hands of the bride and groom-to-be as a blessing. The stain is a rich red-brown colour when used on the body, and usually colours hair red. Clearly the Lover appreciated his coloured and tattooed Beloved.

Nard or Spikenard is named only in *Song of Songs*. It is obtained from several plants, most commonly from a member of the Valerian family, *Nardostachys jatamansi*. It is an oil used as a perfume or as incense. It was probably the incense used in the Temple of Solomon and later, on the specialized incense altar (see the many references in Exodus and Leviticus, when the Hebrew Bible uses the word *HaKetoret*, 'the incense'). Much later it was part of the beginning of the story of the births of John the Baptist and his cousin Jesus Christ:

ST LUKE
CH. 1, vv. 8-9

*Once when Zechariah's division was on duty and
he was serving as priest before God, he was chosen
by lot, according to the custom of the priesthood, to go
into the temple of the Lord to burn incense.*

DILL

The Angel Gabriel came to him as he stood by the incense altar and told him his wife, Elizabeth, would bear a son. Six months later the same angel visited Mary and told her she had been chosen by God to bear His son, Jesus.

Nard was also a medicine to fight insomnia, birth difficulties and other minor ailments. Saffron was another luxury item, this time derived from the Saffron Crocus *Crocus sativus*, a cultivated form of the wild Crocus of the Mediterranean region. The lilac to mauve flowers are harvested in the autumn. It is such an expensive spice, used in flavouring and the orange-yellow colouring of food, because the Saffron powder is produced from the freshly picked flowers – but only the red stigmas or 'threads' from the centres of the flowers are collected for the best product. Today 450 g (1 lb) of dry Saffron needs a harvest of 50,000–75,000 flowers, or a kilogram (around 2 lb) requires 110,000–170,000! Cultivation on this scale is very expensive. The Beloved was a wealthy woman to have a farm and workers to be able to grow enough Crocuses for her Saffron.

SAFFRON CROCUS

Calamus or Sweet Flag *Acorus calamus* probably originated in Asia but is found widely in Europe. It is a wetland plant, whose scented leaves and even more strongly scented rhizomes (the plant's rootstock), have been used for many centuries in medicines, perfumes and as substitutes for ginger, cinnamon and nutmeg – which may be some of the spices suggested by the Lover. We know it was used as early as 1300 BC in Ancient Egypt. Its mention by the Lover may be most closely linked to its long having been a symbol of love, which was very much at the heart of his song. Elsewhere, it is mentioned once each in Isaiah (ch. 43, v. 24) and Jeremiah (ch. 6, v. 20), and is then most likely referring to it as an ingredient in an anointing oil. Jeremiah clearly lists it as an imported item into Israel, not a local product – but 'from a distant land' - again suggesting it was a luxury item for the Beloved to be growing.

The writer of Proverbs also mentions 'myrrh, aloes and cinnamon' together in one quotation (Proverbs ch. 7, v. 17). Writing in the role of an adulteress, the author describes covering her bed with coloured linens from Egypt, which she then perfumes with these three products. Myrrh is a gum commonly harvested from a small thorny tree, *Commiphora myrrha*, native to the southern Arabian peninsula and north-east Africa. A related species, *Commiphora gileadensis*, grows in the eastern Mediterranean. It is clearly mentioned by the prophet Jeremiah as part of his cry of anguish:

"Is there no balm in Gilead?" (Jeremiah 8, v. 22). Gilead was the territory on the east side of the River Jordan just to the north of the Dead Sea. It was an important source of spices and was in the list of goods carried by the merchants who bought Joseph from his brothers (Genesis ch. 37, v. 25). Myrrh was used as an ingredient in the embalming of bodies in Ancient Egypt, in sacred incense in the Temple, in the holy anointing oil (Esther ch. 2, v. 12) and as a drug to dull the senses. To Christians the last use is most famously mentioned as one of the three gifts given to Jesus at His birth by the three

Broad Bean Flowers and Seed Pod

wise men (St Matthew ch. 2, v. 11), and then in the drink offered to Him by the Roman soldiers at His crucifixion (St Mark ch. 15, v. 23).

Aloes are succulent plants of many species, widespread across Africa, the Middle East and Asia. *Aloe vera* is the species most commonly used in herbal medicines, to produce a soothing ointment, to treat wounds and in the making of soap, and is still used in the pharmaceutical industry today. The ancient Greeks and Romans, for example, used it to treat wounds. After the death of Jesus, Joseph of Arimethea took the body with Pilate's permission to inter it in the tomb that Joseph had prepared for himself, as was the custom. Nicodemus, who was another secret follower of Jesus and was the same man who had visited Jesus by night (St John ch. 3), accompanied Joseph:

ST JOHN
CH. 19, VV. 39-40

Nicodemus brought a mixture of myrrh and aloes, about seventy-five pounds [about 34 kilos]. *Taking Jesus' body, the two of them wrapped it, with the spices, in strips of linen. This was in accordance with Jewish burial customs.*

The quantity of spices was unusually large, such as might have been used at a royal funeral, of King Asa of Judah (2 Chronicles ch. 16, v. 14), for example. It shows how important the two men thought Jesus was. To them He *was* indeed the King, but as Jesus said to Pilate at His trial, His kingdom was not of this world. Pilate did not understand that nor did the soldiers who made the crown of thorns and yelled sarcastically, "Hail king of the Jews!" When He was put on the cross, the notice 'Jesus of Nazareth, King of the Jews' was nailed to the cross too. The Jewish authorities protested, saying it should say that He claimed to be the king. But Pilate refused to change it. Jews today are still waiting for the Messiah, the king; and around the world today there are many people of other faiths who scorn Christ as did the Roman soldiers, kill those who do follow Him, and destroy Christian churches. Christ's gospel of love is hard to believe in, and hard to stay true to when one does believe – witness the fact that Joseph and Nicodemus worked secretly, and not many years afterwards Stephen was stoned to death because he refused to deny Christ.

The spices tell a mixed story of human love of others, love of one's own appearance, our love of tasty food, and our care for the sick and injured and dead. From small seeds and flowers great things grow when nurtured with the love of God.

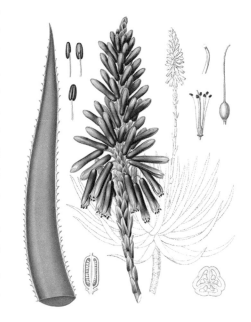

ALOE

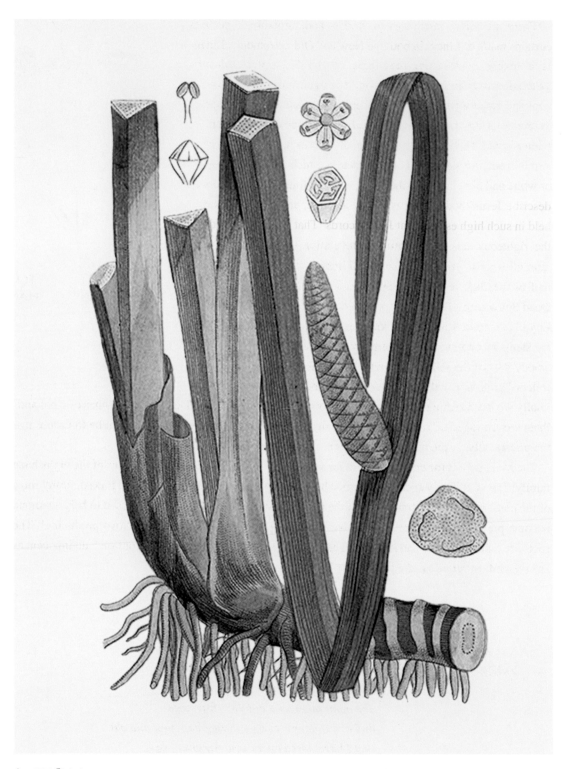

Sweet Flag

There are many references to clothes and furnishings, such as curtains made of Linen, in both the New and Old Testaments. Linen is a textile made from the fibres of the Flax plant *Linum usitatissimum*. Garments made from it are good for keeping you cool and staying fresh in hot weather. When Pharaoh put Joseph in charge of the country 'He dressed him in robes of fine linen' (Genesis ch. 41, v. 42); the palace of Xerxes who reigned in Persia from 486–465 BC, had a garden in which were 'hangings of white and blue linen' (Esther ch.1, v. 6); all four Gospel writers describe Jesus' body being wrapped in linen; and the cloth was held in such high esteem that John records 'That fine linen stands for the righteous acts of the saints' (Revelation ch. 19, v. 9). Egyptian mummies were wrapped in linen; and linen cloth was found in the first half of the 20th century in Cave One during the excavations of the Dead Sea scrolls. The earliest recorded production of the cloth is in Egyptian records 4,000 years old. The fibres are obtained by soaking the stems in water to loosen them (this is called 'retting'); next the woody part of the stalk is removed by crushing the plant between rollers ('scutching'); then the fibres are combed, spun into yarn and finally woven. One unusual law (Deuteronomy ch. 22, v. 11) states 'Do not wear clothes of wool and linen woven together'). The law doesn't say why but commentators think it might be to ensure that the priests' robes were all pure linen cloth.

FLAX
PLANT

The Mandrake is the common name for several species of the genus *Mandragora* of the nightshade family. The most likely species described below is *Mandragora officinarum*. The forked, fleshy roots of the plant resemble the lower part of the human body and, if eaten, were supposed to help a woman become pregnant. The Hebrew word for it is *doo-dah'-ee*, meaning literally 'love producing'. The roots are hallucinogenic and narcotic. The Beloved we have just read about is not such an innocent as she seemed earlier when she says to her Lover:

Let us go early to the vineyards
to see if the vines have budded,
if their blossoms have opened,

SONG OF SONGS *and if the pomegranates are in bloom –*
CH. 7, VV. 12-13 *there I will give you my love.*
The mandrakes send out their fragrance
and at our door is every delicacy, both new and old,
that I have stored up for you, my lover.

MANDRAKE

Finally we will consider the sacred plant Hyssop, a member of the mint family, and native to southern Europe and the Middle East. It is used as an aromatic herb and as a medicine. In English in botanical references, *hyssop* is usually the plant *Hyssopus officinalis,* but in the Bible stories of religious rituals, another mint with a straight stalk and a hairy surface to its leaves which hold liquids well is considered a better contender – *Origanum syriacum* or *O. maru*, popularly known as Bible Hyssop. Its first mention is a dramatic one in the time leading up to the Israelites leaving Egypt:

EXODUS
CH. 12, VV. 21-23

Then Moses summoned all the elders of Israel and said to them, "Go at once and select the animals for your families and slaughter the Passover lamb. Take a bunch of hyssop, dip it into the blood in the basin and put some of the blood on the top and on both sides of the door frame. Not one of you shall go out of the door of his house until morning. When the Lord goes through the land to strike down the Egyptians he will see the blood on the top and sides of the door frame and will pass over the doorway, and he will not permit the destroyer to enter your houses and strike you down.

Later, a law was made which said that a person who had suffered an infectious disease could be cleansed by sprinkling him or her with blood which was on the tip of a sprig of Hyssop (Leviticus ch. 14, vv. 1–9). The writer of *The letter to the Hebrews* firmly reminds his readers of this use of the Hyssop stick when he writes about the blood of Christ, and declares

HEBREWS
CH. 9, V. 9

In fact, the law requires that nearly everything be cleansed with blood, and without the shedding of blood there is no forgiveness.

Today Christians celebrate this in the sacrament of Holy Communion, the Eucharist, when they eat the bread and drink the wine, the body and blood of Christ, in remembrance of His taking the sins of the world on His shoulders (1 Corinthians ch. 11, vv. 23–26).

HYSSOP

CHAPTER IV
MAMMALS

There are many references to mammals in the Bible, both wild and domesticated. Repeatedly they are mentioned not just for their own sake to illustrate the wildness of the Holy Land and the success of the farmers, but also to illustrate God's relationship with his chosen people.

The wild animals were there long before Man, so we will start with them. We begin with a symbol of sovereignty, strength, brutality and courage, the Lion *Panthera leo,* the second largest wild cat. Wild lions exist now only in sub-Saharan Africa and a small population in India. In the time covered by the Bible they were widespread across North Africa and the Middle East, and were kept in menageries by the Romans. In more modern times a subspecies, the Barbary Lion, was kept at the Tower of London, and was the model for the lion sculptures in Trafalgar Square, London.

In scripture we first meet a Lion when Jacob is reunited with all his sons after they had met Joseph in Egypt, and blesses them one by one with words especially chosen for each young man:

GENESIS
CH. 49, VV. 8-9

*"Gather round so that I can tell you
what will happen to you in days to come...
Judah, your brothers will praise you;
Your hand will be on the neck of your enemies;
your father's sons will bow down to you.
You are a lion's cub, O Judah;
you return from the prey, my son.
Like a lion he crouches and lies down,
like a lioness – who dares to rouse him?*

Judah and Joseph were given the longest blessings, which was fitting because their tribes became the leading tribes of southern and northern Israel respectively. Judah was the fourth son born to Jacob, but his three older siblings had forfeited their right to lead, so Judah was next in line and his name

LION CUBS

became the name of the whole of southern Israel in the time of the divided kingdom, roughly 900–600 BC. The Lion of Judah is still used sometimes to describe Israel today.

There is a close link between the Lion of Judah and Ethiopia's history. A 13th century document asserts descent from a retinue of Israelites who returned with Makeda, the Queen of Sheba from her visit to King Solomon in Jerusalem by whom she had conceived the Solomonic dynasty's founder Menelik I. As Solomon was of the tribe of Judah, his son Menelik I would continue the line, which according to Ethiopian history was passed directly down from king to king until Emperor Haile Selassie (ostensibly the 225th king from King David) was deposed in 1974.

A striking sign of the Lion being used as a symbol of sovereignty was shown in the description of Solomon's throne at the time of the visit of the Queen of Sheba:

  1 KINGS
CH. 10, VV. 18-20

Then the king made a great throne inlaid with ivory and overlaid with fine gold. The throne had six steps, and its back had a rounded top. On both sides of the seat were armrests, with a lion standing beside each of them. Twelve lions stood on the six steps, one at either end of each step. Nothing like it had ever been made for any other kingdom.

LION AND LIONESS

Solomon wanted visitors to be in no doubt about how splendid a king he was. What a difference from the way that Jesus taught His disciples to live! They were not to worry about their lives, what to eat, what to wear:

ST LUKE
CH. 12, VV. 27-28

Consider [he said] *how the lilies grow. They do not labour or spin. Yet I tell you, not even Solomon in all his splendour was dressed like one of these. If that is how God clothes the grass of the field, which is here today, and tomorrow is thrown into the fire, how much more will he clothe you, O you of little faith!*

Solomon reigned for 40 years, from 970–930 BC, and on his death the splendour of his rule was broken up and the kingdom was divided in two, Israel in the north and Judah in the south. Enmity remained between the two and neighbouring countries, till the fall of Judah to the Assyrians in 722 BC.

It's interesting that of the roughly 140 references to the lion in the Bible there are only eight in the New Testament, and half of them are in Book of Revelation. The Israelites in the early days of writing repeatedly recorded God as a God of might, but Jesus, and the writers in the New Testament, shift the emphasis to a God of love, so it is not surprising that the lions took a back seat.

From the many references to the lion's roar, it is clear that it was a familiar sound to many people in Old Testament times, and which led to such descriptions as:

PSALM
104, V. 21

The lions roar for their prey

PROVERBS
CH. 20, V. 2

A king's wrath is like the roar of a lion

ISAIAH
CH. 5, V. 29

[The soldiers'] *roar is like that of a lion, they roar like young lions; they growl as they seize their prey and carry it off with no-one to rescue.*

1 PETER
CH. 5, V. 8

Your enemy the devil prowls around like a roaring lion looking for someone to devour

REVELATION
CH. 10, V. 3

[The mighty angel] *gave a loud shout like the roar of a lion*

Travel in Palestine in ancient times was dangerous because of wild beasts. Isaiah several times speaks of the danger of being attacked by lions, but in one more hopeful comment says God will bring faithful people to a glorious, safe land:

ISAIAH
CH. 35, vv. 8-10

And a highway will be there;
it will be called the Way of Holiness.
The unclean will not journey on it;
it will be for those who walk in that Way;
wicked fools will not go about on it.
No lion will be there,
nor will any ferocious beast get up on it;
they will not be found there.
Only the redeemed will walk there,
and the ransomed of the Lord will return.

There were certain roads, such as those between temples, which were specially built to be safe. They were open only to those who were ceremonially pure. Jesus opened up the Way to everyone when, centuries later, He declared to Thomas and the disciples that He was the Way. If you know Christ, you know the way to meet God the Father; an artificial, specially designed road is not needed.

The prophet Micah, a contemporary of Isaiah, wrote words of encouragement in the second half of the eighth century BC, picturing the Israelites as having the strength of a lion:

MICAH
CH. 5, v. 8

The remnant of Jacob will be among the nations,
in the midst of many peoples,
like a lion among the beasts of the forest...
You hand will be lifted up in triumph over your enemies,
and all your foes will be destroyed.

But Ezekiel, prophesying in the late 500s BC, using the same animal as a symbol, wrote a long, detailed lament, which portrayed the kings in violent ways:

What a lioness was your mother among the lions!
She lay down among them and reared her cubs.
She brought up one of her cubs,
and he became a strong lion.
He learned to tear the prey and he became a man-eater.

EZEKIEL
CH. 19, VV. 1-9

The nations heard about him,
and he was trapped in their pit.
They led him with hooks to the land of Egypt.
When she saw her hope unfulfilled,
her expectation gone,
she took another of her cubs
and made him a strong lion.
He prowled among the lions,
for he was now a strong lion.
He learned to tear the prey
and he became a man-eater.
He broke down their strongholds
and devastated their towns.
The land and all who were in it
were terrified by his roaring.
Then the nations came against him,
those from regions round about.
They spread their net for him,
and he was trapped in their pit.
With hooks they pulled him into a cage
and brought him to the king of Babylon.
They put him in prison,
so his roar was heard no longer
on the mountains of Israel.

The kings can be named because Ezekiel dates his story better than any other Old Testament writer – the first lion was Jehoahaz, a bloodthirsty tyrant; and the next was Jehoiachin or Zedekiah, both of whom were taken captive to Babylon after the fall of Jerusalem in 586 BC.

Probably the best known story in the Bible about lions is the tale of Daniel in the lions' den. Darius, the king of Persia appointed 120 governors to rule his kingdom, and three administrators over them. Daniel was one of the three, and soon made his colleagues jealous because of his exceptional qualities of leadership. So they tricked the king into arresting Daniel for praying openly to God, and not worshipping the king. Darius had to obey his own law and Daniel was put in the lions' den, to be savagely executed:

DANIEL
CH. 6, VV. 10-23

At first light of dawn, the king got up and hurried to the lions' den. When he came near the den, he called to Daniel in an anguished voice, "Daniel, servant of the living God, has your God, whom you serve continually, been able to rescue you from the lions?"

Daniel answered, "O king, live for ever! My God sent his angel, and he shut the mouths of the lions. They have not hurt me, because I was found innocent in his sight. Nor have I done any wrong before you, O king."

The king was overjoyed and gave orders to lift Daniel out of the den. And when Daniel was lifted from the den, no wound was found on him, because he had trusted in his God.

The king executed the men who had falsely accused Daniel. He prospered after that in the reigns of Darius and Cyrus, and then in the first year of the reign of Belshazzar king of Babylon (550BC) he had a strange dream – which was a vision of four beasts. The first we have just read about, the Lion, which later is interpreted as the kingdom of Babylonia.

SYRIAN BEAR

Bear with Cubs

> *Then came before me a second beast, which looked*
> *like a bear. It was raised up on one of its sides, and it*
> *had three ribs in its mouth between its teeth. It was told,*
> *"Get up and eat your fill of flesh."*

DANIEL
CH. 7, VV. 3-7

> *After that, I looked, and there before me was another*
> *beast, one that looked like a leopard. ... and it was*
> *given authority to rule.*
> *After that, in my vision at night I looked, and there*
> *before me was a fourth beast – terrifying and*
> *frightening and powerful.*

It is widely believed that Daniel was foretelling the power of the Medes and Persians (the bear), the Greeks (the Leopard) and the Romans (the terrifying beast). Compared with the lion, there are few references in the Bible to the bear, but they are striking nevertheless. The bear was so well known that there is the saying 'Better to meet a bear robbed of her cubs than a fool in his folly' (Proverbs ch. 17, v. 12). The species concerned is the Syrian Bear *Ursus syriacus*, which is regarded by other authorities as just a subspecies of the Brown Bear *Ursus arctos* of Europe and Asia. In Bible lands it is no longer in Israel, Lebanon and Syria, because of habitat destruction and hunting. We read of it very dramatically in one story:

2 KINGS
CH. 2, VV. 23-24

> *From there [Jericho] Elisha went up to Bethel.*
> *As he was walking along the road, some youths came*
> *out of the town and jeered at him. "Go on up, you baldhead!"*
> *they said. "Go on up, you baldhead!" He turned round,*
> *looked at them and called down a curse on them in the*
> *name of the Lord. Then two bears came out of the*
> *woods and mauled forty-two of the youths.*

Baldness was uncommon among Jewish men, and in contrast a good head of hair was taken as a sign of strength. By insulting Elisha the youths were expressing the city's contempt for God's messenger who was Elijah's successor, and who had taken on the task of trying to bring the king of Samaria back to God-fearing ways. Bears can certainly be very dangerous, as young David had explained, especially the she-bear when she has cubs. Not that Elisha was saying that these Bears were going to attack everyone in Israel, but they were a symbol of the message that the whole nation would suffer somehow if it continued to disobey God's commands. During the summer the bears live in the mountains, have a cave as a den, and feed mainly on roots and vegetables. In the winter they descend to the valleys and are known to raid gardens, so it seems most likely that this story about Elisha took place in the winter.

The Israelites had a limited knowledge of astronomy compared with today, but they were in awe of the constellations. We know this thanks to Job who replies to his impatient visitor Bildad, saying:

≥ JOB
CH. 9, VV. 7-9
(AND CH. 38, V. 32)

[God] speaks to the sun and it does not shine;
he seals off the light of the stars.
He alone stretches out the heavens
and treads on the waves of the sea.
He is the Maker of the Great Bear and Orion,
the Pleiades and the constellations of the south. ≥

It was well known in Roman mythology, was known to the Jews as The Bear, and we still refer to the Great Bear and its importance in being able to lead viewers and navigators to find the Pole Star, pointing travellers to the north.

Daniel's other visionary animal is well attested in all translations, the Leopard *Panthera pardus*, one of the so-called 'five big cats' – lion, tiger, leopard, snow leopard and jaguar. It is found across Africa, the Middle East to Siberia. There are only eight references in the Bible to the beast, seven of which are in the Old Testament. We can well believe it was well known when we read:

≥ JEREMIAH
CH. 13, V. 23

Can the Ethiopean change his skin
or the leopard his spots?
Neither can you do good
who are accustomed to doing evil. ≥

LEOPARD

The prophet knew well the appearance of the black-skinned African and the wonderfully spotted coat of the Leopard, and expected his readers and listeners to know them too, and so understand his message to the sinful King of Israel and his people. The prophet had another warning too:

 JEREMIAH
CH. 5, V. 6

Therefore a lion from the forest will attack them,
a wolf from the desert will ravage them,
a leopard will lie in wait near their towns
to tear to pieces any who venture out …

Very clearly the prophet describes the ambushing tactic of a hunting leopard. Notwithstanding the threat these animals posed to people, the prophet Isaiah looked forward to a time when Israel would be freed by a Messiah, a saviour, of the house of David. Then:

ISAIAH
CH. 11, VV. 6-7

The wolf will live with the lamb,
the leopard will lie down with the goat,
the calf and the lion and the yearling together,
And a little child will lead them.
The cow will feed with the bear,
their young will lie down together,
and the lion will eat straw like the ox.

Centuries later it still seems an impossible picture that these predators would ever live side by side with farm animals. But Isaiah wanted people then (and us now) to believe that the Messiah would bring that sort of peace. Christians today believe that Jesus is that peacemaker, and expects us to love as He loved, and so be living proof of what He said: "Blessed are the peacemakers for they will be called the sons of God" (St Matthew ch. 5, v. 9).

John's vision in Revelation speaks of a beast rising out of the sea, which many commentators today take to mean that John was secretly referring to the Roman empire. The beast combines three of the characteristics of the beasts in Daniel's vision (Daniel ch. 7):

REVELATION
CH. 13, V. 2

The beast I saw resembled a leopard, but had feet like those of a bear and a mouth like that of a lion. The dragon gave the beast his power and his throne and great authority.

All readers of St John in those early days of the Christian church would have been in no doubt about the dangerous times they were living in, as described by the image of three major predators being linked together. Today the leopard is in danger of extinction in Israel, with a population of no more than single figures, in the Judean desert and the Negev.

Wolves still stir up emotional reactions from people as they did in Old Testament times. Attempts to reintroduce Wolves to parts of North America and Europe, including Britain, where they used to live, are being made, but meet stiff resistance from farmers who think of the Wolf in just the same way as their Palestinian forbears. Isaiah named the Wolf, and from its mention there and elsewhere in the Old and New Testaments, it is clear the Wolf *Canis lupus* is meant, the largest member of the dog family. Wolves are widespread across the whole northern hemisphere, and are divided into many subspecies which vary subtly in size, colour and behaviour. The form called the Arabian Wolf *Canis lupus arabs* is still found in southern Palestine. It is a small subspecies, mostly adapted to being a desert dweller, and weighing only about a quarter as much as a Eurasian Wolf. The species used to be the world's most widely distributed mammal, but now lives in barely two-thirds of its former range, reduced in distribution and numbers by deliberate persecution because of its preying on livestock.

We first meet it in Jacob's blessings again, where it seems more like a curse than a blessing:

⊲ GENESIS
CH. 49, V. 27

Benjamin is a ravenous wolf;
in the morning he devours the prey,
in the evening he divides the plunder. ⊳

WOLVES ATTACKING WILD OX

One can only wonder what Benjamin thought of his father's words, but suffice it to say that some of Benjamin's descendants were a fierce, savage people (see Judges chh. 19–21). The Wolf's ferocity was also referred to centuries later by Jesus when in His early preaching He told His hearers:

ST MATTHEW   *Watch out for false prophets. They come to you in sheep's*
CH. 7, V. 15   *clothing, but inwardly they are ferocious wolves.*

In both Matthew's and Luke's gospels Jesus is recorded as telling His disciples when He sends out the twelve to preach and to heal, that

ST MATTHEW   *I am sending you out like sheep among wolves. Therefore be*
CH. 10, V. 16   *as shrewd as snakes and as innocent as doves.*

Jesus was well aware that many of the Jews, especially those in authority, would vilify their message and even harm them. The wolf's cruelty and greed were so reviled in Bible times that bad government officials (Ezekiel ch. 22) and judges (Zephaniah ch. 3) are described as being like wolves.

The Hebrew word for 'wolf' is 'ze'ebh' and the word is even used as a man's name. One of the Midianites defeated by Gideon (Judges ch. 4) was called Zeeb, a fact recalled in Psalm 83, and the men who are the elders of Jerusalem are described as 'evening wolves, who leave nothing for the morning' (Zephaniah ch. 3, v. 3). Wolves certainly prowled around sheep folds at night. Jeremiah speaks of its night-time hunting (if you believe the *Authorised Version*), but it is a desert hunter according to modern translations. The fierceness of the Wolf demands faithful protection of the flock by the shepherd, as we shall read later (see p. 95).

Wolves were clearly a very likely danger to flocks in those days and were therefore a familiar symbol used by preachers and prophets to illustrate bad ways of living, to reinforce the message in order to turn people to God's good way.

There are just one or two more predatory mammals mentioned in the Bible. A teacher of the law, probably a scribe or a Pharisee, came to Jesus one day and said:

ST MATTHEW   *"Teacher, I will follow you wherever you go."*
CH. 8, V. 20   *Jesus replied, "Foxes have holes and birds of the air have*
nests, but the Son of man has nowhere to lay his head."

Jesus' enigmatic answer gets no reply that we know of from the other teacher. The man probably very quickly realised that the cost of following Jesus was going to be hard; he had to be prepared to 'live rough'. The other references to foxes are all in the Old Testament and are all referred to by the same Hebrew word 'shu'al' which also means 'jackal'. Is it a fox or a jackal which the Lover sings of in his song?

SONG OF SONGS
CH. 2, V. 15

Catch for us the foxes,
the little foxes
that ruin the vineyards,
our vineyards that are in bloom. ✑

The Syrian Fox (or Jackal) is very likely the subspecies of the Golden Jackal *Canis aureus syriacus* which is native to the eastern Mediterranean, from Lebanon to Tripoli. It does live in a burrow which suits Jesus' description, and is destructive in vineyards as the Lover says, because it is very fond of ripe grapes. So, naturally, the Lover would like his Beloved to order her workers to catch the animals before they ruin her vines which are now in bloom.

The Hyena, another predator, is mentioned by Isaiah in his prophecy about the destruction of Babylon:

✑ **ISAIAH**
CH. 13, V. 22

Hyenas will howl in her strongholds,
jackals in her luxurious palaces. ✑

This is a rare mention in the Bible of what must be the Striped Hyena *Hyaena hyaena* found across northern Africa, the Middle East, Arabia and India. Isaiah maybe mentions it again when he describes the downfall of Israel's enemies, the Edomites:

✑ **ISAIAH**
CH. 34, VV. 13-14

Thorns will overrun her citadels,
nettles and brambles her strongholds.
She will become the haunt for jackals,
a home for owls.
Desert creatures will meet with hyenas … ✑

The *NIV* is alone in translating the original as 'hyenas' instead of 'wild animals'. It is primarily a scavenger of carrion, and mainly hunts at night. Its feeding habits alone would make it a detestable creature to the Israelites, so it is an ideal animal to highlight the descriptions of desolation and death.

Foxes are used as a different symbol, as in Jesus' response to the Pharisees:

✑ **ST LUKE**
CH. 13, VV. 31-32

At that time [on His way to Jerusalem] *some Pharisees*
came to Jesus and said to him, "Leave this place and go
somewhere else. Herod wants to kill you."
He replied, "Go tell that fox, 'I will drive out demons
and heal people today and tomorrow, and on the third day
I will reach my goal.' " ✑

Striped Hyena

The fox clearly had a reputation then of being crafty, just as we think of it, in the story of Little Red Riding Hood, for example. Or is the word better translated as it is in this characterisation?

EZEKIEL
CH. 13, V. 4
THE GOOD NEWS BIBLE

People of Israel, your prophets are as useless as foxes living among the ruins of a city.

We are back to a translation problem. The *New International Version* says these prophets are 'jackals among ruins'. *The Living Bible* says 'foxes'. *The Message* prefers 'jackals'. There are foxes in Palestine, a subspecies of the Red Fox we know, *Vulpes vulpes*, or the Syrian Fox *V. thaleb*. The Hebrew word 'shu'al' is used for both. It is an interesting puzzle, but whether jackal or fox, the divine message of each comment is the same – people and countries who were not going with God on life's journey were sure to suffer God's wrath, be they crafty or useless.

So much for predators. What about Man? What prey may he have hunted? This is carefully described in the list of food that God told the Israelites they may eat, the so-called 'clean food' – which implies that they were the hunters too, because several species listed are wild rather than farm animals. Mankind is as much a part of the natural history of the lands of the Bible as the wild animals. Lions and wolves, described above, preyed on a variety of species. The prey Man may have hunted is in that list:

RED FOX

DEUTERONOMY
CH. 14, VV. 4-5

These are the animals you may eat: the ox, the sheep,
the goat, the deer, the gazelle, the roe deer, the wild goat,
the ibex, the antelope, and the mountain sheep.

Although the commentaries admit that some of the animals and birds in the lists in Deuteronomy and Leviticus of clean and unclean foods are not possible to identify for certain, modern translations are in general agreement with the list above of the so-called 'clean animals'. These are the animals which a Jew today will say are 'kosher'. According to the laws of the Torah, the only types of meat that may be eaten are cattle and game that both have 'cloven hooves' and 'chew the cud.' If an animal species fulfils only one of these conditions (for example the pig, which has split hooves but does not chew the cud, or the camel, which chews the cud, but does not have split hooves), then its meat may not be eaten. Examples of kosher animals in this category are bulls, cows, calves, sheep, lambs, goats, and antelopes. According to the laws of the Torah, before it is eaten, a kosher species must be killed by a ritual slaughterer. Since Jewish Law prohibits causing any pain to animals, the slaughtering has to be effected in such a way that unconsciousness is instantaneous and death occurs almost as quickly.

Flocks and herds of animals were used to describe the wealth of the patriarchs and early heroes. As before, it is sometimes not easy to interpret exactly which animals are intended. In the Old Testament the Hebrew word 'eleph' becomes 'cattle' in our translations. In such cases it is often to be understood as a 'portmanteau' term, which includes sheep, goats, oxen and donkeys. In fact there are 13 different Hebrew words in the Old Testament which are at one time or another translated as 'cattle'. In the *NIV* we often find simply 'cattle' where *Good News* expands the description to 'sheep and goats'. The two animals together make a very reasonable translation because when the Israelites were in the desert for 40 years they depended on their flocks of these two animals for food and clothing. This essential wealth which was vital to survival is aptly recorded in the words of the writer of this proverb:

PROVERBS
CH. 27, VV. 25-27

When the hay is removed and new growth appears
and the grass from the hills is gathered in,
the lambs will provide you with clothing,
and the goats with the price of a field.
You will have plenty of goats' milk
to feed you and your family
and to nourish your servant girls.

Although Hebrew does have a word for fresh milk, as soon as it was put in a goatskin bottle it went sour, so both goat's and sheep's milk was used as the main ingredient in the production of cheese and

IBEX

butter. These get just one reference each in the Old Testament; 'churning the milk' to produce butter (Proverbs 30, v. 33) was a sensible thing to do; and when the Israelites were fighting the Philistines, Jesse said to his son David:

1 SAMUEL
CH. 17, V. 18

"Take along these ten cheeses to the commander of their unit. See how your brothers are and bring back some assurance from them."

For Jesse to have such a considerable store of cheese indicates how busy the women must have been. Cheese-making usually began in May, and goats' cheese was considered best.

Jesus spoke about God's final judgement of everyone in a seemingly mysterious simile:

ST MATTHEW
CH. 25, VV. 31-33

When the Son of man comes in his glory, and all the angels with him, he will sit on his throne in heavenly glory. All nations will be gathered before him, and he will separate the people one from another as a shepherd separates the sheep from the goats. He will put the sheep on the right and the goats on his left.

In many cultures the left has been associated with misfortune, disaster, evil even. Goats were less valuable than sheep so it was natural for Jesus to put them on the left. The Romans had a word in Latin for this belief – *sinistra* – from which we get our word 'sinister'. Jesus was not denigrating the value of goats but was simply using them to illustrate the way that on Judgement Day God will separate good people from bad. But 'separating the sheep from the goats' has become an English saying to describe any two sorts of people who cannot be left together.

Notwithstanding the apparently less than favourable value of the goat, the animal was an important part of sacrificial worship. A man who had confessed to an unintentional sin (that is, he had done what is forbidden by God), could pay for that sin by sacrificing as a burnt offering a 'female goat without defect' (Leviticus ch. 4, v. 28).

Apart from mentioning the herds of goats, the Bible also has several references to 'wild goats'. For example, young David fled from King Saul's anger and hid among the rocks of the wild goats (1 Samuel ch. 24, v. 2); and God asked Job if he knew when the wild goats have their young (Job ch. 39, v. 1). These may refer to the Ibex.

The Domestic Goat *Capra hircus* is derived from the Wild Goat *C. aegargus*. Beginning about 10,000–11,000 years ago, Neolithic farmers in the Near East began keeping small herds of goats for their milk and meat, for their dung which they used as fuel, as well as for their hair, bone, skin and sinew which they used as materials for clothing and building. Archaeological evidence has revealed that domestication took place about 10,000 years ago in the valley of the River Euphrates in Turkey, and in the mountains of Iran.

Scholars believe cattle were domesticated over 10,500 years ago, probably from several wild species at the same time in various parts of the world. Wild Cattle in the Fertile Crescent of the Near East were the likely ancestors of the cows the Hebrews kept, and are today considered to be a subspecies of the Aurochs *Bos primigenius* of Europe, Asia and Africa which survived in the wild in Europe till the mid 1600s. The Romans used them in the arena in bullfights.

It was not until the Israelites were settled in Palestine that they had enough pasture that was suitable for rearing herds of cows. During their time in the wilderness, many such animals would not have survived. But each tribe did have some cattle, as we shall read below. Oxen were vitally important in the agricultural economy of the ancient Jews. They ploughed the land, trod out the corn and pulled carts. Herds of cattle never became as numerous as sheep and goats, but nevertheless when the people came together to celebrate the Feast of the Tabernacles and bring the Ark of the Covenant to the temple:

 2 CHRONICLES
CH. 5, V. 6

King Solomon and the entire assembly of Israel that had gathered about him were before the Ark, were sacrificing so many sheep and cattle that they could not be recorded or counted. ≋

Although 'cattle' may mean several kinds of domestic animal, clearly many cows were there. Evidence that good pasture was needed is found in the psalm praising God's bounty which says 'He makes grass grow for the cattle' (Psalm 104, v. 14). The majority of the farmers were not wealthy and may have had only one or two oxen, not to rear for beef, but as draught animals. A farmer needed to plough his land and an ox (or two yoked together) would pull the one-bladed plough, steered by the ploughman. Depending on the crops he could grow, the farmer had to plough his ground at least twice a year. The normal length of a furrow according to Varro Reatinus (116–27 BC) in his *Reum Rusticarum Libri Tres* (Three Books on Agriculture), was 37 m (120 ft), anything longer was too much of a strain for the oxen, who were rested at the turning point. They were often kept in a stall at night as recorded by the prophet Habakkuk (ch. 3, v. 17). Oxen were also trained to pull a cart. A line in a psalm is not able to be translated with certainty, but one version says 'our oxen will draw heavy loads' (Psalm 144, v. 14), and at the Dedication of the Tabernacle in the desert we learn more of their importance to the farmers, when the leaders of the twelve tribes:

 NUMBERS
CH. 7, V. 3

Brought as their gifts before the lord six covered carts and twelve oxen – an ox from each leader and a cart from every two. ≋

Wild Goat

These were shared out among the leaders who had particular work to do. In the New Testament one of the most memorable references to oxen is Jesus telling guests at a Pharisee's house the Parable of The Great Banquet. One man refused the invitation to come, saying "I have just bought five yoke of oxen, and I am on my way to try them out. Please excuse me." (St Luke ch. 14, v. 19). The whole parable was Jesus' way of emphasizing that everyone will eat at the feast in the kingdom of God, not just important guests.

The Old Testament has a reference to training cattle, which backs up the story by showing that the man did need to see if the oxen were right for the job. After the Philistines had fought the Israelites and captured the Ark of the Covenant they were inflicted with an outbreak of tumours. The people believed the Ark had caused the pestilence, and their priests said the Ark had to be returned, together with a guilt offering of gold representing the tumours and the places affected:

1 SAMUEL
CH. 6, VV. 7-8

Now then, get a new cart ready, with two cows that have been calved and have never been yoked. Hitch the cows to the cart, but take the calves away and pen them up. Take the Ark of the Lord and put it on the cart ... and the gold objects.

Needless to say, the Israelites rejoiced.

AUROCHS

Agriculture was an ancient lifestyle in the Middle East, going back 10,000 years or more. Seasonal work in that era is famously recorded in the Gezer Calendar (see p. 50). By the time of Jesus we have more documentary evidence, and it is clear that farmers were still growing crops and keeping animals as was described by Varro and by Cato the Elder (234–149 BC), author of *De Agri Cultura* (On Agriculture), comprising 162 chapters on home and estate management. The farmer's year was timed according to three festivals, in particular associated with harvesting barley, wheat and summer fruits, but animal husbandry was a continuous activity.

In Genesis we read that Abel kept flocks, and these were largely sheep and goats. Sheep are believed to be one of the earliest animals to be domesticated, mostly from the wild Mouflon *Ovis orientalis* in Mesopotamia, for their fleece, meat and milk. Sheep were so rich a part of settled life it was only natural that they should figure in the mythology and religion of the people of the ancient Middle East and the Mediterranean region. Shrines of the ancient Egyptians from about 8000 BC contain rams' skulls; the fertility god, Amun, was depicted in the form of a sheep. Later, it is not surprising that we find in the Bible that sheep play an important part in the lives of Abraham, Isaac, Jacob, Moses and King David (as a boy), who were all shepherds. These are the so-called Fat-tailed Sheep, a general type of domestic sheep known for their distinctive large tails and hindquarters. Fat-tailed Sheep breeds comprise approximately 25 per cent of the world population, and are commonly found in northern parts of Africa, the Middle East, and Asia as far as western China. The earliest record of this sheep variety is found in ancient Uruk and Ur (c. 2400 BC) on stone vessels and mosaics. Another early reference is found in the Bible (Leviticus ch. 3, v. 9), where a sacrificial offering is described which includes the tail fat of sheep.

One of the earliest Bible stories which tells of sheep and shepherding is the well-known tale of Joseph and his coat of many colours, made famous in *Joseph and the Amazing Technicolor Dreamcoat*, an operetta from the 1970s with lyrics by Tim Rice and music by Andrew Lloyd Webber. Jacob and his family were suffering from famine and went to Egypt where they had heard there was still wheat. The family did not know at that time that the youngest son, Joseph, who had been sold to be a slave by his jealous brothers, was a leader in Pharaoh's court:

GENESIS
CH. 47, vv. 1–3

Joseph went and told Pharaoh, "My father and brothers, with their flocks and herds and everything they own, have come from the land of Canaan and are now in Goshen." He chose five of his brothers and presented them before Pharaoh. Pharaoh said to the brothers, "What is your occupation?" "Your servants are shepherds, just as our fathers were."

There was no hesitation in saying what they did, and how long the family had followed that livelihood. Throughout Old and New Testament times pastoral work was of the greatest importance, and shepherds were kept in high esteem – witness the fact that shepherds were chosen to be among the first to hear about the birth of Christ:

ST LUKE
CH. 2, vv. 9-11

And there were shepherds living out in the fields nearby, keeping watch over their flocks at night. An angel of the Lord appeared to them, and the glory of the Lord shone around them, and they were terrified. But the angel said to them, "Do not be afraid. I bring you good news of great joy that will be for all people. Today in the town of David [Bethlehem] a saviour has been born to you; he is Christ the Lord.

MOUFLON

St Luke tells that they hurried off and found Mary and Joseph and the baby lying in a manger. It is one of those stories that we take for granted as we hear it every Christmas. I cannot help wondering when we realise the importance of their work and how hard it was, which we shall read about below, did they all go? was one left behind to guard the sheep? did the angels stand guard?

There are several references stating just how large the flocks were that some families had. The conquered King Mesha of Moab had to pay an annual tribute to the king of Israel of 100,000 lambs and the wool of 100,000 rams; Job owned 7,000 sheep; and at the dedication of his temple, Solomon ordered the sacrifice of 22,000 head of cattle and 120,000 sheep and goats (2 Kings ch. 3, Job ch. 1; 2 Chronicles ch. 7). Even if these figures are exaggerated as some believe, the Psalmist praised God for His bounty by saying that the fields were covered with sheep. He must have been telling what he had seen, not just once but repeatedly (Psalm 65). There can be no doubt that sheep were a very noticeable part of daily life.

The most striking thing about the way a Palestinian shepherd worked was that he led the flock and they followed him, unlike an English shepherd who drives his flock, and keeps control with the help of one of more dogs. The sheep follow because they recognise the voice of their shepherd, or even the sound of the flute which some of them played. When David went to join King Saul's court, he was introduced as one who could play the harp. Did he play when he was shepherding? God and the leaders of Israel are symbolically called shepherds, as stated by the prophets for example:

JEREMIAH
CH. 31, V. 10

Hear the word of the Lord, O nations;
proclaim it in distant coastlands:
he who scattered Israel will gather them
and will watch over his flock like a shepherd.

God will save his people in the same way that a shepherd will gather together his flock which has wandered in different directions. The image of the shepherd was used too, to warn leaders of the errors of their ways:

EZEKIEL
CH. 34, VV. 1-5 & 10-16

The word of the Lord came to me: "Son of man,
prophesy against the shepherds of Israel; prophesy and
say to them: 'This is what the Sovereign Lord says: Woe
to the shepherds of Israel who only take care of themselves!
Should not shepherds take care of the flock? You eat the
curds, clothe yourselves with the wool and slaughter the
choice animals, but you do not take care of the flock. You
have not strengthened the weak or healed the sick or bound
up the injured. You have not brought back the strays or

searched for the lost. You have ruled them harshly and brutally. So they were scattered because there was no shepherd, and when they were scattered they became food for all the wild animals ... I am against [these] shepherds and will hold them accountable for my flock. ≈

This long chapter is a detailed account of a shepherd's work and responsibilities, and the life of the sheep. It becomes an analogy for the way all those who provide leadership – kings and their officials, and prophets and priests – and all the people they govern, should live safely and with justice together. A shepherd was indispensable. Over them all they should acknowledge that God, their Sovereign Lord, is the Great Shepherd of the Sheep. Isaiah put it very clearly:

≈ ISAIAH
CH. 40, V. 11

He tends his flock like a shepherd:
He gathers the lambs in his arms
and carries them close to his heart;
He gently leads those that have young. ≈

He clearly describes God as a personal, caring God, not as a remote, commanding figure. The Psalmist wrote about it too in Psalm 23, with words that are often read at funerals: 'The Lord is my

SHEPHERDS

shepherd … your rod and your staff, they comfort me'. This symbolic use of sheep and shepherds was carried through seamlessly centuries later into the early Christian faith. It is not surprising, therefore, that Jesus, who was brought up as a faithful Jew and well versed in the scriptures and agricultural life of His time, refers to Himself as the Good Shepherd, and the people He lives with and teaches as His sheep:

ST JOHN
CH. 10, vv. 14-15

I am the Good Shepherd; I know my sheep and my sheep know me – just as the Father knows me and I know the Father – and I lay down my life for the sheep.

At the end of the day the shepherd would lead his flock to the sheepfold. This was a circular area surrounded by a stone wall, open to the sky, and having only one entrance. It was well known to Jesus who used it in a figure of speech to further illustrate the meaning of His saying He was a shepherd to the people:

ST JOHN
CH. 10, v. 7, 9

I am the gate for the sheep... Whoever enters through me will be saved.

This imagery of shepherd and sheep is very movingly part of the life of Simon Peter. After the Resurrection, Peter and four other disciples were fishing and Jesus helped them bring in a very large catch of 153 fish. They all had a picnic meal of bread and fish cooked over an open fire.

ST JOHN
CH. 21, v. 15

When they had finished eating, Jesus said to Simon Peter, "Simon, son of John, do you truly love me more than these?" "Yes, Lord," he said, "you know that I love you." Jesus said, "Feed my lambs."

Very significantly Jesus tells Peter three times to "feed my sheep". The message sinks home, because he had denied Jesus three times. Years later when Peter writes his first letter in the mid 60s AD to the Christians scattered in Asia Minor, he tells the elders of these churches:

1 PETER
CH. 5, vv. 2-4

Be shepherds of God's flock that is under your care, serving as overseers – not because you must, but because you are willing, as God wants you to be … eager to serve, not lording it over those entrusted to you, but being examples to the flock. And when the Chief Shepherd appears, you will receive the crown of glory that will never fade away.

Peter faithfully continued to shepherd God's flock till his martyrdom in the late 60s AD in the reign of the Emperor Nero.

The wealth of a family was closely allied to the flocks they had. It was natural therefore, that in Judaism when people needed to show their gratitude for the wealth they enjoyed, they did it by sacrificial worship. A sheep or a lamb was sacrificed by burning it on the altar at festival times, and in rituals associated with particular burnt offerings – the Peace or Fellowship Offering, and the Guilt Offering. When Nehemiah was given permission by King Artaxerxes to return to Jerusalem to rebuild the city in about 440 BC he started with the walls, and 'Eliashib the high priest and his fellow priests went to work and rebuilt the Sheep Gate' (Nehemiah ch. 3, v. 1). It was significant that it was the high priest who started work here because this was the route by which lambs and sheep were brought to the temple for sacrifice. It was the only gate which was sanctified because it had such an important spiritual significance.

For the Jews, the animal sacrifice was a positive, practical way of showing that they were closer to God. Christians earnestly believe that Jesus made the ultimate sacrifice by His death on the cross, and so they believe their sins are indeed forgiven. Sheep and lambs are used in the Bible as symbols of Christ: 'The next day John saw Jesus coming toward him and said, "Look, the Lamb of God, who takes away the sin of the world!"' (John ch. 1, vv. 29–30).

Humans are believed to have begun to domesticate the Wild Horse *Equus ferus* in about 4000 BC in central Asia, and scholars believe they were completely domesticated by 2000 BC, with the full range of harness we are familiar with today. There is only one form of truly wild horse left today, the rare subspecies, Przewalski's Horse *E. f. przewalskii*, to be found in the steppes of Central Asia. Horses in the Bible, however, are most often war-horses, not farm animals. They were used for pulling chariots and as mounts for cavalry soldiers. We first meet them in Egypt where they were introduced as a weapon in the 16th century BC. A splendid relief carving of Pharaoh Ramses II riding with archers at the Battle of Kadesh is on the wall of Abu Simnel. The defeat of the Egyptians during the flight of the Jews from Egypt is recorded in the Song of Moses and Miriam:

> *I will sing to the Lord,*
> *for he is highly exalted.*
>
> EXODUS   *The horse and rider*
> CH. 15, VV. 1 & 4   *he has hurled into the sea …*
> *Pharaoh's chariots and his army*
> *he has hurled into the sea.*

When the Israelites had settled in the Promised Land, God told them that 'The king, moreover, must not acquire great numbers of horses for himself, or make the people return to Egypt to get more of them' (Deuteronomy ch. 17, v. 16), but

> *Solomon accumulated chariots and horses; he had fourteen hundred chariots and twelve thousand horses, which he kept in the chariot cities and also with him in Jerusalem ... Solomon's horses were imported from Egypt and from Kue.*

2 CHRONICLES
CH. 1, VV. 14 & 16

Kue is probably the area we now call Cilicia, the south-eastern corner of modern Turkey. As a strong leader Solomon could not resist adding the 'ultimate deterrent' to his army, despite God's command. Isaiah bewailed this form of authority when he cried:

> *Woe to those who go down to Egypt for help,*
> *who rely on horses,*
> *who trust in the multitude of their chariots*
> *and in the strength of their horsemen,*
> *but do not look to the Holy One of Israel,*
> *or seek help from the Lord.*

ISAIAH
CH. 31, V. 1

The Old Testament is full of galloping horses, the rattle of wheels, powerful war-horses and prancing horses. But the New Testament has only three mentions of these animals. When Paul was arrested by the Sanhedrin and was being taken for trial by Roman law (because he was a Roman citizen), the local commander:

> *called two of his centurions and ordered them, "Get ready a detachment of two hundred soldiers, seventy horsemen and two hundred spearmen to go to Caesarea tonight. Provide mounts for Paul so that he may be taken safely to Governor Felix."*

ACTS OF
THE APOSTLES
CH. 23, VV. 23-24

When James in his letter is explaining how much damage we can do by speaking, using our tongue, he says,

When we put bits into the mouths of horses to make them obey us, we can turn the whole animal ... but no man can tame the tongue. It is a restless evil, full of deadly poison. ⤳

The third reference is in the Bible's last book, Revelation. John the author, has a vision and sees the Lamb (i.e. Jesus) open one of seven seals on a scroll which would certainly have been thought of as a legal document, because seven was considered a number denoting completeness. In the vision the breaking of the seals is a symbolic action which precedes the blowing of trumpets, and pouring out of bowls, and all are the foretelling in heaven of dreadful events which will the happen on earth. First and most dramatically, come war-horses, and their riders have become known as the Four Horsemen of the Apocalypse. These are described in Revelation ch. 6, vv. 1–8. The four riders are on white, red, black and pale green horses. They are the harbingers of conquest, war, famine and death, leading up to the Last Judgement. Some authorities believe the vision applied only to first century life under the Romans. Others think a future tribulation will kill many. Yet a third interpretation sees the horsemen as representing current events, even particular beliefs, so the red horse represents Communism, the white horse Catholicism, the black horse Capitalism and the pale green horse Islam. The discussion will run for a long time yet. John may be recalling the horsemen on red, brown and white horses seen by Zechariah in a vision (Zechariah ch. 1, vv. 8–11), but they were reporting to God that the 'whole

DROMEDARY

world was at rest and in peace', not the fact that nation will rise against nation as was heralded by Jesus (St Mark ch. 13).

Like horses, camels are also referred to as a means of transport. These camels are the single-humped Dromedary *Camelus dromedaries*, which were first domesticated about 4,000 years ago in Arabia. They are easier to train than cattle, so besides carrying goods they can be used to pull a cart or a plough. An adult Dromedary can carry 160–290 kg (352–640 lb) for 24 km (15 miles) a day for a long time, thanks to its adaptation to desert life. They were, therefore, an essential part of commercial transport.

When Joseph's jealous brothers were wondering what to do with him after they had kidnapped him:

≫ GENESIS
CH. 37, V. 25

As they sat down to eat their meal, they looked up and saw a caravan of Ishmaelites coming from Gilead. Their camels were loaded with spices, balm and myrrh, and they were on their way to take them down to Egypt. ≫

These men were all descendants of Abraham, and they perfectly illustrate the length of the journeys undertaken by these caravans. Gilead was the hill country east of the Jordan, between the Sea of Galilee and the Dead Sea, and their route would have been south-west via the Negev and the Wilderness of Shur, 350–450 km (200–300 miles). Their spices are all described in Chapter 3.

We assume that the Three Wise Men or Magi of the Christmas story arrived on camels. They almost certainly were well-educated astrologers from Persia or southern Arabia, so the only way in those days to travel such a long distance, several hundred miles, was by camel. They are mentioned only in St Matthew's Gospel (chapter 2), which says nothing about how they arrived, only that they came 'from the east'.

Camels were also valuable providers of milk. Camels' milk when mixed with three parts of water was considered a good, nourishing drink. Their hair was woven and made into clothing – 'John wore clothing made of camel's hair, with a leather belt round his waist' (St Mark ch. 1, v. 6) – and their skin was used to make tents St Paul stayed and worked with Aquila and his wife Priscilla when he was in Corinth, 'because he was a tentmaker as they were' (Acts ch. 18, v. 3).

The account of the Israelites' return to Jerusalem from exile in about 537 BC is recorded in the books of Ezra and Nehemiah. There is a detailed list of the types of people and their animals:

≫ EZRA
CH. 2, vv. 64-67

The whole company numbered 42,360, besides their 7,337 menservants and maidservants; and they also had 200 men and women singers [who sang at social events such as weddings, as distinct from the temple singers who were all male]. *They had 736 horses, 245 mules, 435 camels and 6,720 donkeys.* ≫

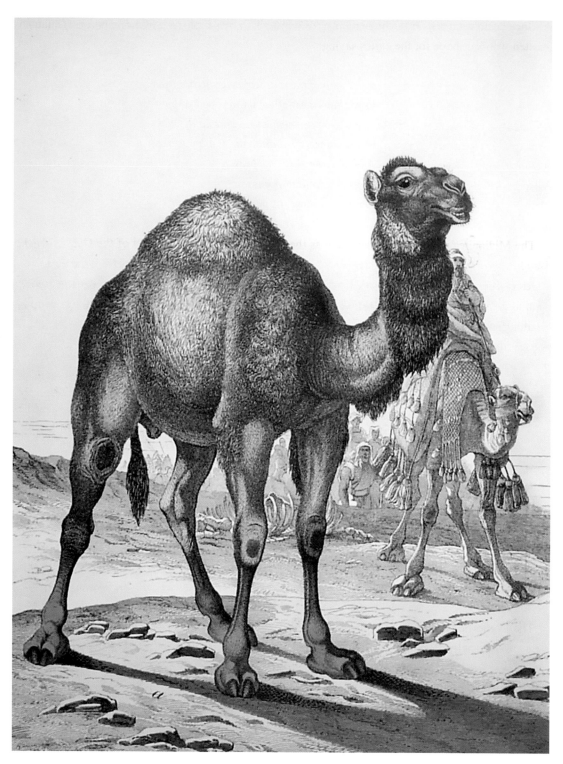

Camel and Camel Train

Once again we discover the huge number of livestock these people owned. Long before, Isaiah had written words of hope for the exiles saying:

 ISAIAH
CH. 60, vv. 5-6

the wealth of the seas will be brought to you,
to you the riches of the nations will come.
Herds of camels will cover your land,
young camels of Midian and Ephah.
And all from Sheba will come,
bearing gold and incense
and proclaiming the praise of the Lord.

The Midianites roamed the deserts across the Jordan and south to the east of the Gulf of Aqaba. Isaiah is referring to caravans of camels led by merchants bringing luxury goods from the east.

Jesus named the camel in one of His talks to the disciples after He had failed to persuade a young man to give up his rich life and so find his treasure in heaven. The young man could not give up his wealth. Jesus told the disciples who were there:

ST MARK
CH. 10, vv. 23-25

"How hard it is for the rich to enter the kingdom of God!"
The disciples were amazed at his words. But Jesus said again,
"Children, how hard it is to enter the kingdom of God!
It is easier for a camel to go through the eye of a needle than
for a rich man to enter the kingdom of God."

Jesus stresses the rich person's problem by contrasting the largest animal in Palestine with the smallest opening. Very many people today have the same problem, having a greater love of material possessions than for eternal life.

There are three times more references in the whole Bible to donkeys than to camels. The reason is clear: donkeys were part of daily life for the many farmers and folk like Joseph and Mary as a beast of burden, whereas camels were the chosen form of transport for commercial business, carrying goods, often great distances, for far fewer merchants, or for the wealthy to ride, such as the Wise Men.

The Domestic Donkey *Equus asinus* was bred from the wild African Ass *E. africanus* about 6,000 years ago in ancient Egypt. For example, Tutankhamun's tomb has a mural depicting a wild ass hunt. The beast was well adapted for the semi-desert life of these people, to carry the possessions of the early Israelites on their travels.

Wild animals were still well known when the Israelites were settled in the Promised Land, as is remarked by the Psalmist:

[God] makes springs pour water into the ravines;
it flows between the mountains.
They give water to the beasts of the field;
the wild donkeys quench their thirst.

We first read of donkeys in the story of the early part of Abram's life in Egypt where he had travelled to avoid the famine in Canaan. He settled and 'acquired sheep and cattle, male and female donkeys, manservants and maidservants, and camels' (Genesis ch. 12, v.16). A family's wealth was measured in the number and quality of their livestock, so clearly Abram had become a wealthy man.

Livestock were valuable, so were a legitimate spoil of war, and when Moses led the Israelites to victory over the Midianites the soldiers took 61,000 donkeys – and 675,000 sheep, 72,000 cattle and 32,000 young women! (Numbers ch. 31, vv. 32–33).

In Numbers chapter 22 is a splendid story about a donkey. Balaam, a pagan prophet and seer, was an expert in interpreting the future from the actions of or remains of animals – animal divination. God

DONKEY

had forbidden him to curse the Israelites, which Balak, king of the Moabites, was commanding him to do. But God does allow him to go to the Israelites, so long as Balaam follows God's instructions. He sets off intending, however, to follow his own plan and be paid a reward by Balak. He rides off on his donkey, which suddenly turns off the road down a narrow path, because it sees an angel of the Lord. Balaam beats the donkey, which says,

NUMBERS
CH. 22, VV. 28-30

"What have I done to you to make you
beat me these three times?"
Balaam answered the donkey, "You have made
a fool of me!"....
The donkey said to Balaam, "Am I not your own donkey,
which you have always ridden, to this day?
Have I been in the habit of doing this to you?"
"No, "he said.

The proverbially dumb animal was able to see the angel but Balaam, whose skills were well known abroad, was now spiritually blind. Balaam then sees the angel, and admits he has sinned.

There are several references to riding a donkey with a saddle. No doubt Mary had one for her three-day journey from Nazareth to Bethlehem, made so that they could register themselves as required by the census ordered by Caesar Augustus. That journey, and the one of well over 160 km (100 miles) shortly afterwards to Egypt to escape Herod's order to kill newborn boys, are mentioned only in St Matthew's gospel. Neither says anything about a donkey! That they travelled that way is conjecture.

There is no doubt, however, about the way that Jesus rode into Jerusalem at the end of His earthly life:

ST MATTHEW
CH. 21, VV. 1-5
&
ZECHARIAH
CH. 9, V. 9

As they approached Jerusalem and came to Bethphage on the
Mount of Olives, Jesus sent two disciples, saying to them,
"Go to the village ahead of you, and at once you will find a
donkey tied there, with her colt by her. Untie them and bring
them to me. If anyone says anything to you, tell him that the
Lord needs them, and he will send them right away."
This took place to fulfil what was spoken through the prophet:
"Say to the daughter of Zion, 'See your king comes to you,
gentle and riding on a donkey, on a colt, the foal of a donkey'"

Evidence for this donkey is in all four Gospels. For long the donkey had symbolised humility, peace and royalty of the house of David – to which Jesus did belong. He deliberately rode this way to emphasise His being a peaceful saviour of the Jews, not the leader of a strong military force to oust the Romans, which is what the Zionists claimed.

A trip to Palestine or North Africa today is like time travel into the past. You will see, as I have, men riding donkeys from the fields towards home, donkeys pulling small carts, and donkeys laden with bundles of wood or animal fodder.

Gazelles are mentioned in Proverbs ch. 6, v. 5 where we read a warning against folly:

Free yourself, like a gazelle, from the hand of the hunter.

Isaiah, too, speaks of the 'hunted gazelle' (ch. 13, v. 14). But the Beloved sees the animal very differently:

SONG OF SONGS
CH. 2, VV. 8-9

Listen! My Lover!
Look! Here he comes,
leaping across the mountains,
bounding over the hills.
My lover is like a gazelle or a young stag.

In the early years of the Christian church Peter travelled in Judea and visited Joppa north-west of Jerusalem.

ACTS OF
THE APOSTLES
CH. 9, V. 36

In Joppa there was a disciple named Tabitha
(which when translated, is Dorcas), who was
always doing good, and helping the poor.

Both names, the one Aramaic, the other Greek, mean 'gazelle'. The disciple was named after the gazelle which lived in the hills around her home. The gazelle is an antelope belonging to a family of even-toed ruminants, which is exactly what the law demanded – the Israelites could eat 'any animal that has a split hoof divided in two'. There are more than 20 species of gazelle, all belonging to Asia and Africa. The species still found in Palestine is the Dorcas Gazelle *Gazella dorcas*. It stands 1.8 m (2 ft) high at the shoulders. Both sexes have horns, which may be 30 cm (12 in) long. It is an attractive animal, an ideal one to describe the Lover. The general coloration is tawny, but it is creamy white below and on the rump, and has a narrow white line from above the eye to the nostril. The Dorcas Gazelle is found singly or in small groups on the interior plains and the uplands, such as the single one

Dorcas Gazelles

I saw on a ridge by the road from Jerusalem to Jericho. A herd, when alarmed, makes off with great rapidity over the roughest country, as is suggested in the description of how well one of David's men could run – 'as fleet-footed as a wild gazelle' (2 Samuel ch. 2, v. 18). The skin is used for floor coverings, pouches or shoes, and the flesh is eaten, though not highly esteemed.

The Roe Deer *Capreolus capreolus* is found widely across Europe and Asia. It is a little taller than the Gazelle, has a lovely red-brown hide, a white rump patch, and only the male (buck) has short, erect antlers. It is a woodland species, and in the Mediterranean region is mostly found in the hills and mountains. Many a person right up to modern times, feeling oppressed by troubles like the Psalmist, has exclaimed

> PSALM
> 42, vv. 1-2
>
> *As the deer pants for streams of water,*
> *so my soul pants for you, O God.*
> *My soul thirsts for God, for the living God.*

The vividness of the image suggests the writer had seen such a hunted deer, in desperate need of life-giving water, even as he needs reviving by God. Elsewhere the few references to deer speak of their speed and sure-footedness (2 Samuel ch. 22, v. 34; Habakkuk ch. 3, v. 19), and graceful appearance (Proverbs ch. 5, v. 19).

Readers brought up on the words of *The Authorized* or *King James* version of the Bible which was read in churches for nigh on 400 years before being superseded by modern translations, would have read or heard of a strange animal mentioned in Leviticus and Proverbs:

> PROVERBS
> CH. 30, V. 26
>
> *The conies are but a feeble people, yet make*
> *their houses in the rocks.*

The conies (singular 'cony' or 'coney'), according to the writer, were one of 'four things which are little upon the earth, but they are exceeding wise' (v. 24). 'Cony' is an archaic English word for 'rabbit'. Besides the list of 'clean animals', the Israelites were given a list of 'unclean animals' which they should not eat:

> LEVITICUS
> CH. 11, vv. 3 & 5-6
>
> *The coney, though it chews the cud, does not have a split hoof;*
> *it is unclean for you. The rabbit, though it chews the cud,*
> *does not have a split hoof; it is unclean for you.*

The early translators did not really understand the Hebrew word 'shapham' because they had no knowledge of the animal the writer of Leviticus was referring to – the Rock Hyrax *Procavia capensis*, (or Coney), which is mostly a sub-Saharan species, but whose distribution spreads north through

Israel to Lebanon and Syria. Furthermore, they do not chew the cud, but make a distinctive grunting sound and move their jaws at the same time, thus giving the impression to the ancient Jews that they were chewing the cud. Family groups live in underground dens in rocky territories, feed less than a 100 m (100 yd) from the entrance on plant foods, and fall prey to Leopards, snakes and eagles, if the one on watch fails to see the predator in time to sound a warning.

One of the keys to understanding the Bible is our comprehension of the symbolism that is hidden in the naming of God's creatures. We have read about many of them from the largest to the smallest, all with a rich tale to tell – from strength to weakness, from war to peace, from human greed to God's countless gifts of love.

CHAPTER V

BIRDS

I srael is well known as being one of the best places in the world to witness bird migration. Millions of migrants from hundreds of species pass through Israel twice a year, making it one of the world's busiest and most impressive flyways.

Israel's remarkable list of birds is due to its geography and varied habitats. From the snowy peaks of Mt Hermon at over 2,200 m (7,200 ft) above sea level via the Great Rift Valley and down to the Dead Sea depression, the lowest place on Earth 440 m (1,440 ft) below sea level, each habitat hosts unique birds and wildlife.

Israel is the northern limit of distribution for African bird species like Little Green Bee-eater *Merops orientalis* and Namaqua Dove *Oena capensis*, the southern limit for European species like finches, jays and others, and the western limit of distribution for fascinating Asian species like kingfishers, bulbuls, babblers and more.

The flora and fauna in Israel are so varied that visitors are sometimes amazed at the biodiversity to be found in quite small areas. For birdwatchers the figures speak for themselves, 540 species of bird have been recorded. For comparison the checklist for the United States of America is over 950 species, and 580 in Great Britain. This is quite impressive, especially when we remember that Israel is a relatively small country, under 480 km long by 110 km wide (300 by 70 miles), smaller than England.

The Bible records that on the fifth day of Creation God said:

> *"Let the waters teem with living creatures, and let birds fly above the earth across the expanse of the sky." So God created the great creatures of the sea and every living and moving thing with which the water teems, according to their kinds, and every winged bird according to its kind. And God saw that it was good. God blessed them and said, "Be fruitful and increase in number and fill the water in the seas, and let the birds increase on the earth."*

GENESIS
CH. 1, vv. 20-22

The birds have indeed been fruitful and increased. Today they are found on every continent, in every habitat, from the snowy wastes of Antarctica to the steaming hot rainforests around the Equator. Scientists believe there are about 11,000 species; new ones are discovered every year in remote places; and others which were once thought to be a single species with slight variations in plumage or behaviour, are now known to be separate species thanks to the study of their DNA.

Bee-eaters at nesting colony

The birds of the story of creation gave the Old Testament writers in particular great inspiration to describe the world in which the people lived. A thousand or more years after Genesis was written an anonymous author wrote *The Wisdom of Solomon*. It is not in the Hebrew Bible, but was accepted by Jerome for his Latin translation, which we know as the Vulgate Bible. It is still part of the canon for Roman Catholics and many orthodox groups, but is only accepted in the Apocrypha in the Protestant Bible. The writer was fascinated by the passing of time and the lack of signs to show which creatures had lived before in that place:

WISDOM OF
SOLOMON
CH. 5, V. 11

When a bird has flown through the air, there is no token of her way to be found, but the light air being beaten with the stroke of her wings, and parted with the violent noise and motion of them; it is passed through, and afterwards there is no sign in the air to be found of where she went.

We still say today, "Oh, how time flies!" But for most of us our minds then rush on to think of other things, and we do not allow our imaginations to explore the thought of flying time as the writer did centuries ago.

Bird sounds greatly affected the writers in ancient Palestine. Much of that which we can read in the Bible today concerns the mournful sounds of the dove and the owl, two birds we shall consider more carefully later. But the sound of birdsong is recorded too, lifting the spirits of writer and reader:

SONG OF SONGS
CH. 2, VV. 11-12

See! The winter is past;
the rains are over and gone.
Flowers appear on the earth;
the season of singing has come,
the cooing of doves
is heard in our land.

Whether this love song was written by King Solomon as is traditionally thought, or was simply about him, does not spoil the poem's romantic imagery which is timeless and as understandable in our culture today as it was in the Middle East hundreds of years before Christ. Apart from the oft-repeated cooing of doves, this and another reference (see opposite) are the only ones to the songs as opposed to the calls of birds generally. The most likely songster is the Yellow-vented Bulbul *Pycnonotus xanthopygos*, which is found throughout the Middle East. It is a resident in Israel and is very common in some parts of the country, a bird of gardens, plantations and oases, wherever trees and bushes are to found. So, it is even common in villages and towns. Its bold, repetitive song proclaiming that it has a territory, is especially to be heard in spring, in March and April, starting before sunrise. I have

YELLOW-VENTED BULBUL

listened to one in April singing from a television aerial overlooking the Pool of Bethesda, in the old part of Jerusalem.

But of all the writings in the Old and New Testaments the composition which most closely records the author's wonder at God's Creation is Psalm 104. This is not a repeat description of the order of creation but a hymn to the Creator praising the wonders that the writer sees before him – the heavens, the seas, the valleys, the mountains are all there, richly filled with plants, animals, food, and:

> *The birds of the air nest by the waters;*
> *they sing among the branches ...*
> *The trees of the Lord are well watered,*
> *the Cedars of Lebanon that he planted.*
> *There the birds make their nests;*
> *the stork has its home in the pine trees.*
>
> — PSALM 104, v.v 12, 16, 17

Apart from the words which describe the glory of creation, the carefully controlled structure of the psalm is a further sign of the psalmist's sense of wonder and desire to praise the Lord. The main part of the psalm begins with the heavens (v. 1) and ends with the seas (v. 26), the two elements which, in his experience, hold the earth where he sees the splendours of creation. As we stop to consider the world around us – springtime primroses, a rainbow, snow-capped hills, the first Swallow of the new year – we can well understand why the psalmist sang God's praises like 'the birds of the air'.

It is worth putting the Hebrews' knowledge of birds into a worldwide context so that we can get a feel of how accurate their observations were.

Historical evidence from times past in Egypt where the Israelites were in bondage for many years, shows that their masters were aware of the

PYGMY CORMORANT

birds that lived around them. Accurate paintings of many species can be seen on the walls in the ancient tombs.

In the Bible there are about 300 references to birds. They range from the mention of 'bird/birds' as a general term, to very specific descriptions. The former in the original Hebrew uses one of two general words. The word 'oph' is for birds and other flying creatures, like insects. It is used in the Genesis story. Another general word is 'tsippor', which usually denotes game birds, like the 'bird from the snare of the fowler' (Proverbs ch. 6, v. 5), or perching birds, like the one 'alone on a roof' (Psalm 102, v. 7). Moses' wife's name, Zipporah, is derived from this word. In the New Testament Greek there are also two words for birds. Birds in general are 'peteinon' as in St Matthew ch. 6, v. 26: 'Look at the birds of the air; they do not sow or reap or store away in barns'. But 'orneon' refers to flesh-eating birds, like vultures, as in the description of the destruction of Babylon which is 'the haunt of every unclean and detestable bird' (Revelation ch. 18, v. 2).

By the time of Christ the Hebrews, who had lived long in times past among the Egyptians, were now immersed in Greek culture. Use of the Greek language was widespread among the educated. All the writers of the New Testament wrote in Greek, or their scribes did. At that time Palestine was under the control of the Romans, builders of arterial roads and splendid buildings. Early Christians no doubt learnt the skills of the thinkers, artists and writers of their time. Although later than the Bible texts, the Byzantine church at Tabgha near Capernaum illustrates this. The church is on the site where it is believed Christ fed the five thousand (St Matthew ch. 6, St Mark ch. 14). It dates from the fourth century AD, was enlarged in the fifth century, later destroyed by the Muslim invaders, and rediscovered in 1932. The ruins contain the finest floor mosaics in Israel. One depicts a long-necked swan or a goose, and two cormorants in their typical wing-drying pose, together with beasts, fish and ribbons of flowers. The cormorant in the mosaic is a small bird, which looks a reasonable representation of the

Pygmy Cormorant *Phalacrocorax pygmaeus*, not the Great Cormorant *P. carbo,* which is now, and probably always was, an uncommon bird in Palestine. The Pygmy Cormorant used to be a common winter visitor to the Hula wetlands north of Galilee, where it also bred until the 1950s, before the marshes were drained. Another mosaic shows a splendid cock Peacock.

However, unlike the life-like paintings and mosaics, the Biblical text is often not clear as to which species are referred to. As we noted in chapter 4, there were laws (in Deuteronomy and Leviticus) about which animals and birds were 'clean' or 'unclean' (i.e. may be eaten or must not be eaten). And this is explored further below. There is a famous list of unclean birds in Leviticus chapter 11, but translating the species intended is problematical. Here we compare the list of some of the species in the translation by Dr G. R. Driver (*Palestine Exploration Quarterly*, April 1955) with the names in a modern Israeli bird book, published in 1987 by the Society for the Protection of Nature in Israel and Tel Aviv University. Firstly names which show some agreement:

Dr Driver's list 1955		**Checklist of the Birds of Israel 1987**	
Hebrew	*English*	*Hebrew*	*English*
kos	Tawny Owl	kos hehorvot	Little Owl
yanshuph	Screech Owl	yanshuf ezim	Long-eared owl
yanshuf sadot	Short-eared Owl		
dukipheth	Hoopoe	dukhipat	Hoopoe
chasidah	Stork	hasidah	Stork
nesher	Griffon Vulture	nesher	Griffon Vulture
peres	Bearded Vulture	peres	Bearded Vulture
daah	Kite	daya	Kite
oreb	Raven	orev	Raven
anaphah	Heron	anafit or anafat	Heron

Other unclean birds listed in Dr Driver's transliteration can be traced in the modern list, but with a different interpretation:

Dr Driver's list 1955		**Checklist of the Birds of Israel 1987**	
Hebrew	*English*	*Hebrew*	*English*
racham	Osprey	raham	Egyptian Vulture
shalak	Fisher Owl	shalakh	Osprey
tishemeth	Little Owl	tinshemet	Barn (or Screech) Owl
tachmas	Short-eared Owl	tahmass	Nightjar
ayyah	Falcon or buzzard	ayat	Honey buzzard
ozniyyah	Short-toed Eagle	ozniyah	Black Vulture

It is interesting to note that the translation in the *New Revised Standard Version* follows the *Checklist*'s names for 'racham' and 'tachmas'.

Life is from God. Blood is life. So blood is holy, and to eat blood was an offence against God, making the offender ritually 'unclean'. Therefore, the Law said, "Wherever you live, you must not eat the blood of any bird or animal. If anyone eats blood that person must be cut off from his people" (Leviticus ch. 7, vv. 26–27). The meat from which the blood has been drained is 'kosher'. The law is emphasised even more strongly later:

LEVITICUS
CH. 17, vv. 13-16

The Lord said to Moses, "Any Israelite or any alien living among you who hunts any animal or bird that may be eaten must drain out the blood and cover it with earth, because the life of every creature is its blood. That is why I have said to the Israelites, "You must not eat the blood of any creature, because the life of every creature is its blood."

So we are led to the longest, most famous list of birds in the Bible, in Leviticus and repeated in Deuteronomy:

LEVITICUS
CH. 11, vv. 13-19
(NEW INTERNATIONAL VERSION)

These are the birds you are to detest and not eat because they are detestable: the eagle, the vulture, the black vulture, the red kite, any kind of black kite, any kind of raven, the horned owl, the screech owl, the gull, any kind of hawk, the little owl, the cormorant, the great owl, the white owl, the desert owl, the osprey, the stork, any kind of heron, the hoopoe and the bat

SAME REFERENCE IN
THE GOOD NEWS BIBLE

These are the birds you shall regard as vermin, and for this reason they shall not be eaten: the Griffon Vulture (or eagle), the Black Vulture, and the Bearded Vulture (or ossifrage); the kite and every kind of falcon; every kind of crow (or Raven), the Desert-owl, the Short-eared Owl, the Long-eared Owl, and every kind of hawk; the Tawny Owl, the Fisher-owl, and the Screech-owl; the Little Owl, the Horned Owl, the Osprey, the Stork (or Heron), every kind of Cormorant, the Hoopoe and the bat.

Hoopoes

Nineteen species or categories of bird are listed here, in whatever translation you read. There are basically three groups: birds of prey, water birds and two perching birds one of which, the Hoopoe *Upupa epops*, seems on the face of it the odd one out. I am sure the Hebrews knew that a bat was not a bird – but bats do fly, they do eat insects, which are unclean, and so they eat blood and are therefore themselves unclean.

We have already discovered that it is not easy for scholars today to identify some of the birds. I propose here to write in more detail about the birds we can be sure or reasonably sure about, based on the Hebrew name as shown in *Young's Analytical Concordance*, the most commonly used modern translation, help from the modern *Checklist of the Birds of Israel*, and the *NIV* translation.

One bird which all the translations are sure about is the White Stork *Ciconia ciconia*, which is recorded in Jeremiah's prophecy, and which has the most detailed mention of migratory birds to be found in the Bible. He began his ministry in 626 BC, and preached and prophesied for about another 40 years. As with the other prophets, he feels called to declare God's painful awareness of the Jews' sins, their drifting further from God's laws and their lack of repentance. At one point he describes their failure to do the right thing in these terms:

JEREMIAH
CH. 8, V. 7

No-one repents of his wickedness,
saying, "What have I done?"
Each pursues his own course
like a horse charging into battle.
Even the stork in the sky
knows her appointed seasons,
the dove, the swift and the thrush
observe the time of their migration,
But my people do not know
the requirements of the Lord.

All these birds are indeed migrants through or to Israel. The White Stork's migration has been well studied in recent times. The routes of some individuals have been tracked by scientists by means of a tiny radio transmitter fixed to each bird. They breed in Iberia and Eastern Europe; those from Iberia cross to Africa via Gibraltar, those from Eastern Europe cross via the Bosporus into Turkey. The latter then fly on south through the Levant, the Jordan valley in the Holy land, into the Nile Valley, and on to their principal winter quarters from November to February in Kenya, Uganda and eastern Cape Province of South Africa. The two populations have long been known to use these two narrow crossings, thus avoiding a long sea crossing which they dislike. These large birds, whose huge wing-span of 155–165 cm (65–69 in) gives them plenty of 'lift' but not forward speed, do need to gain a considerable height to enable them to have enough air space to progress with slow, steady wing beats

White Stork and Chicks

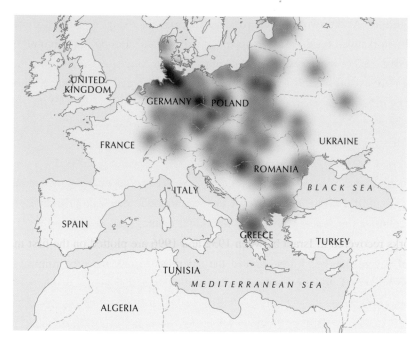

Ringing sites of White Storks recovered in Israel:
darker shades denote greater density

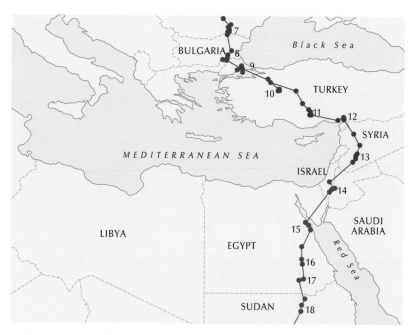

Migration pattern of "Caesar", no. 94555, from
Bulgaria to Sudan: the night locations are numbered

and some gliding. At the two narrow crossings thermals of rising air build up as the day becomes warmer. The storks wait for these thermals to form, then rise, circling in them until they have enough height to move on. For centuries the flocks building up in August at the Bosporus have attracted the attention of tourists and scientists alike. As many as 10,000 may be seen slowly circling, rising high, before setting off south in small groups. The return in spring is the reverse of the one in autumn, but quicker – some birds may be in Russia by early April. Jeremiah wrote 'seasons', and was clearly thinking of the storks in spring *and* autumn.

In 2002 The German Federal Agency for Nature Conservation published a major report (Willem Van den Bossche et al., 2002). They recorded thousands of storks entering modern Israel from Syria, that is from the north-east and east. These flocks stop in the Jordan Valley, in the Bet She'an Valley in particular, south of the Sea of Galilee, for one night or more before moving on. The origin of 317 marked storks recovered in Israel between 1932 and 1996 are plotted on the first map opposite. The reporting of several marked birds enabled the scientists to calculate the minimal daily migration distance the birds had flown between Israel and their breeding area. One started from Israel and reached northern Germany, 3,304 km (2,053 miles) away, in 25 days, an average of 132 km (82 miles) a day. Another reached Poland 2,728 km (1,695 miles) away in 27 days, 101 km (63 miles) a day. One stork, known as Caesar, had a radio transmitter attached and was tracked every day by satellite. This enabled the ornithologists to produce an amazingly detailed map of the bird's journey. It started its migration on 26 August 1994 in northern Germany, spent its tenth night in Turkey, was in Israel for its 14th stopover night, crossed into Egypt the next day and was well into Sudan by day 18. Another known simply by its ring number, 94549, started its migration on the 1st September and was in the Bet She'an Valley on the 14th September, where it stayed for 12 days.

These modern observations are living evidence that it is not surprising that these large, distinctive birds attracted Jeremiah's attention. Although the Psalmist records their nesting in Biblical times in Cypress Trees (Psalm 104), they have rarely nested in The Holy Land in modern times.

Many years later, in historical times, when the Israelites were in exile in Egypt, Moses in particular, at Pharaoh's court, would have become familiar with the way birds were a rich part of religious life there. The Egyptians over the years believed in several gods, whom they represented by birds. The god of knowledge, science and art, Thoth, was represented by the Sacred Ibis *Threskiornis aethiopica*. A carving of the ibis was used as a votive offering, and mummified ibises were buried in many different catacombs as late as the Ptolemaic times (c.350–30 BC). This is a large, noticeable black and white water bird which the Israelites must have been familiar with, but recorded it only once in scripture:

JOB
CH. 38, V. 36
(GOOD NEWS BIBLE)

Who tells the Ibis when the Nile will flood,
or who tells the cock that rain will fall?

The *NIV* translates it differently with no bird names at all, but a footnote does admit the birds are 'possible … and serve as a transition to the next major section' of Job's discourse with God. Perhaps the ibis was not well regarded by the Jews because of its being the representation of an Egyptian god.

The second bird in Jeremiah's list on page 118 is interesting, from biblical and modern birdwatching points of view:

Authorised Version	New English Bible	New Revised Standard Version	New International Version	Good News Bible
stork	stork	stork	stork	stork
turtle (dove)	dove	Turtle Dove	dove	dove
crane	swift	crane	swift	swallow
swallow	wryneck	swallow	thrush	thrush

It is named 'agur' in the Hebrew original. The most modern translations do not translate it 'crane', yet that is still the modern name, 'agur afor' for the Common Crane *Grus grus*, and 'agur hen' for the rarer Demoiselle Crane *Anthropoides virgo*, so why popular translations name other species I do not understand. Like the storks, they are huge, conspicuous birds on migration. They fly with long necks and legs extended, and a wingspan of 220–245 cm (92–102 in.). Travelling in flocks, there are often

COMMON CRANES

hundreds together, flying in a 'V' formation, keeping contact with each other, and announcing their arrival to people on the ground, with loud trumpeting calls. The commonest times to see them are the periods of passage, September to December and March to April. Most of the birds passing through Israel are from northern and eastern Europe; some winter in Israel, but most fly on down the Nile valley to winter in river valleys in Sudan and Ethiopia. These tall, grey birds are surely the species mentioned by Jeremiah. They are conspicuous in flight and on the ground, and Jeremiah and his contemporaries could well have been troubled by them, as modern Israeli farmers are. From overnight roosts, flocks spread out over the fields, and although they do eat wild fruits and seeds, and some small rodents and insects, they are liable to eat vegetables too, especially on the Sabbath when the fields are unattended. Birds which do that are bound to be noticed.

The largest bird in the world today is the Ostrich *Struthio camelus*, and there are six books in the Bible which name it: Leviticus, Deuteronomy, Job, Isaiah, Lamentations and Micah. The most striking one is the detailed account of the life of the Ostrich in Job:

⤳ JOB
CH. 39, VV. 13-18

The wings of the ostrich flap joyfully,
but they cannot compare with the pinions
and feathers of the stork.
She lays her eggs on the ground
and lets them warm in the sand,
unmindful that a foot may crush them,
that some wild animal may trample them.
She treats her young harshly, as if they were not hers;
she cares not that her labour was in vain
for God did not endow her with wisdom
or give a share of good sense.
Yet when she spreads her feathers to run,
she laughs at horse and rider. ⤳

A cock Ostrich is flightless but its white plumes are flapped in the courtship display and as it runs. Ostriches were hunted in ancient times for these feathers as an ancient Assyrian seal shows; they are still much sought after as part of a lady's finery. Ostriches do lay their eggs in a simple depression in the ground where they are in danger of being trampled by wild camels or antelopes. Sometimes two hens lay in one nest, for the cock is polygamous, which may explain why the writer believes the hen doesn't care for her brood. Modern study of this bird has revealed that the male and female do incubate the eggs, and both care for the chicks.

Ostriches used to be widespread in the Middle East, where it was one of six subspecies, the only one outside Africa. It was smaller than the African races. Scientists believe it is one of the most

Ostrich (male in foreground)

primitive birds now in existence. Perhaps, instinctively, the writer of Job realised this, and that is why he portrays it without wisdom. It used to be found in the Syrian and Arabian deserts; in the Negev and Sinai. The proliferation of fire-arms after the First World War, and hunting with motor vehicles were its death knell, and it has been extinct there since the 1930s. In 1973 eighteen chicks were reintroduced into Israel, at the Hai Bar Nature Reserve; it was hoped to breed from them and release them in the Negev. Unlike the problematic identifications of some birds' names, the Hebrew of the Bible terms the Ostrich 'ya-annah', which is very close to the modern 'ya-en'.

The next noticeable group of birds that the Bible records are the birds of prey. Owls, eagles, vultures and falcons are frequently mentioned. As we have already seen in the Leviticus list, the Jews in ancient times were familiar with several species of owl. Only the Lovers heard the songs of birds (see p. 152); it seems that most Israelites had been so oppressed by warring neighbours, and had suffered so long in Exile in the seventh century BC, that certain bird calls were not pleasant, they were sad or ominous, even as today many people fear the call of an owl.

The prophet Micah wrote some years before 700 BC. The kingdom of Judah, the southern part of the Holy Land, was where he lived and observed the people's sinfulness, and lack of service to God. He prophesied that Samaria and Jerusalem would fall and,

 MICAH
CH. 1, VV. 8-9

Because of this I will weep and wail;
I will go about barefoot and naked.
I will howl like a jackal
and moan like an owl.
For her wound is incurable ...

Between 734 and 701 BC the Assyrians did invade, and conquered as far south as Jerusalem, which was spared from destruction. Zephaniah prophesied about 100 years later. He, too, was from Judah and was a descendant of King Hezekiah who was the last King whom Micah knew. Once again he writes that God is about to punish apostate Judah, but the Lord eventually will destroy those who threaten Israel and Judah:

 ZEPHANIAH
CH. 2, V. 14

The desert owl and the screech owl
will roost in her columns.
Their calls will echo through the windows,
rubble will be in the doorways ...

Many modern translations agree that owls will be hooting through the windows but there is little agreement on how to translate the proper names of the birds; and the 'rubble' of the *NIV* is a bird in other translations! We read of the 'horned owl and ruffed bustard' (*New English Bible*), 'tawny owl'

SHORT-EARED OWL

will hoot in the window (*NEB*) and a 'bustard' (*NEB*) or a 'raven' (*New Revised Standard Version*) will 'croak on the threshold'. The *NEB*'s choice of 'bustard' here is curious. Bustards are birds of open grassland; two species are rare in Israel today, through loss of habitat and over-hunting. Even at the best of times a bustard seems an unlikely bird to be seen in the ruined doorway of an Assyrian house. But the mournful sound of an owl, especially the deep '*ooo-hooo, ooo-hooo*' of the Eagle Owl *Bubo bubo*, would undoubtedly have been a sinister sound to emphasize the desolation of a ruined city. These owls are still widespread in Israel. Their call is often made just as they leave a roost at dusk to go hunting. The call is so loud that on a still night it carries for 2–3 km (over a mile). That would certainly have greatly affected impressionable people. It would not have been only Jews who were troubled by owls. In a taunting prophecy against Babylon where the Israelites were in exile, Isaiah cried out in God's name:

> ISAIAH
> CH. 14, VV. 22-33

I will cut off from Babylon her name and survivors,
her offspring and descendants, declares the Lord.
I will turn her into a place for owls
and into a swampland.

Ten species of owl have been recorded in Israel. Mostly they inhabit drier habitat than swampland, but the Barn Owl *Tyto alba* and Short-eared Owl *Asio flammeus* regularly hunt over such ground, even in daylight, so would attract the attention of superstitious people.

Vultures figure prominently in references to death. For example, near the end of His earthly life Jesus, tells His disciples that His second coming will be obvious:

ST MATTHEW
CH. 24, VV. 27-28

For as lightning that comes from the east is visible even in the west, so will be the coming of the Son of Man. Wherever there is a carcass, there the vultures will gather.

Jesus was familiar with the sight of a flock of vultures gathering. They do the dirty work of clearing up after death, at the end of a bloody battle for example. It is fascinating that Jesus uses the image of death that a circling flock of vultures indicates, to show how His *living* reappearance will be just as noticeable.

The Canadian prize-winning natural history writer, Wayne Grady, called the vulture, 'Nature's ghastly gourmet'. Although tourists on African safaris may be keen to see vultures which had been circling in the sky come down to the carcass of a zebra or gnu which lions had killed, many are horrified by what they see close by. But this is the way of vultures. They are Nature's 'dustbin men'. They perform the essential job of clearing up the 'leftovers' after the carnivores have had their fill, or of disposing of the body of a creature that had died naturally. Throughout the centuries men and women who lived close to nature, and many who still do in Africa and India, welcome the flocks of vultures at the village middens. These people are not so squeamish as we are. To us 'ghastly gourmets' they may be, but in many parts of India now there are no vultures, or very few, because of a drug given by vets to cows, which improves the health of the cattle but is a deadly poison to vultures. People are now realising how much they relied on these birds, and that is how the Jews centuries ago must have felt.

They must have served a necessary service in eating the offal from dressed meat and fowls for a family's food, and the entrails discarded from animals and birds at the sacrificial altar. We certainly know the latter because Abram had to drive the vultures away from his altar (Genesis ch. 15). Additionally, when the Israelites were in exile in Egypt, long before they reached the

GRIFFON VULTURE

Promised Land, they would certainly have been aware of vultures, both the Griffon Vulture *Gyps fulvus* and the less powerful Egyptian Vulture *Neophron percnopterus*. Vultures became known as 'Pharaoh's Chickens' after one Pharaoh made it a crime punishable by death to kill a vulture. He realised how important they were in a hot country to keep an area clean and healthy. Moses, in particular, at the palace, would have become familiar with the way these birds were thought of, which is quite different from the western view of them today.

The descriptions in the Bible do not refer to them as useful like Pharaoh did. Rather, when read one after the other, they read like a history of the Israelites breaking God's covenant and commandments, and their punishing defeats at the hands of their enemies. This tragic history started after the death of Solomon when the country he had ruled was divided into two, the northern kingdom of Israel, and the southern kingdom of Judah. The Jews' covenant with God had been described in detail – like an insurance policy – in the book of Deuteronomy. Blessings for Obedience are listed, followed by Curses for Disobedience, in which we read:

DEUTERONOMY
CH. 28, VV. 25-26

The Lord will cause you to be defeated before your enemies. You will come at them from one direction but flee from them in seven, and you will become a thing of horror to all the kingdoms on earth. Your carcasses will be food for all the birds of the air and the beasts of the earth, and there will be no-one to frighten them away.

The last few words are interesting because they do describe one thing which the Jews felt about birds of prey, and that would have especially meant vultures. If no-one was left to scare the birds away, the bodies would be defiled by the birds before there could be any mourning, or the bodies given an honourable burial, according to the prescribed ritual. So the spirit of the dead would not rise to Eternal Life. The curse is exactly what happened in the reign of Jeroboam (930–909 BC). Ahijah the prophet described the king as one who had 'done more evil than all who had lived before him' (1 Kings ch. 14, v. 9). Because of this the prophet told him:

1 KINGS
CH. 14, V. 11

Dogs will eat those belonging to Jeroboam who die in the city, and the birds of the air will feed on those who die in the country. The Lord has spoken.

Those birds of the air were surely ravens, kites – and vultures. It is interesting that the writer says the birds will not be in the city. Red and Black Kites, and Ravens are still well known in many cities. The Raven *Corvus corax*, the all-black and largest of the crows, is known in many countries to be associated with death, either because its harsh call in wild, desolate places foretells it, or because of its being seen feeding on corpses. It is listed in various translations as 'every kind of Raven' or 'the Raven after its kind'. It is clearly an unclean bird, feeding on carrion, insects, reptiles and worms, indeed almost anything edible. The way the lawmakers wrote about it is accurate. In Israel there are in fact three species: the common Raven which is widespread across the northern hemisphere, but rarest of the three in Israel; the Brown-necked Raven *Corvus ruficollis*, which is a common resident in desert areas; and the smallest, the Fan-tailed Raven *C. rhipidurus*, which is quite common in the Dead Sea region, south to Eilat and Sinai in mountainous desert areas. One Spring I watched two of the Fan-taileds displaying over Masada by the Dead Sea, diving and twisting in a splendid series of aerobatic movements, showing clearly their distinctive, heavy-billed, short-tailed silhouette. The writer of *Proverbs* described well the Raven's unclean behaviour:

RAVEN

⤳ PROVERBS
CH. 30, V. 17

The eye that mocks a father,
that scorns obedience to a mother,
will be pecked out by the ravens of the valley ⤳

But intriguingly the Raven is not thought of as 'unclean' in the story of Noah, because the first bird he sent out from the Ark to spy for land was the Raven.

⤳ GENESIS
CH. 8, VV. 6-12

After forty days Noah opened the window he had made in the ark and sent out a Raven, and it kept flying back and forth until the water had dried up from the earth. Then he sent out a dove to see if the water had receded from the surface of the ground. But the dove could find no place to set its feet because there was water over all the surface of the earth; so it returned to Noah in the ark. He reached out his hand and took the dove and brought it back to himself in the ark. He waited seven more days and again sent out the dove from the ark. When the dove returned to him in the evening, there in its beak was a freshly picked olive leaf! Then Noah

knew that the water had receded from the earth. He waited
seven more days and sent the dove out again, but this time
it did not return to him.

The Raven is an interesting choice. It is usually thought of in European folklore – and even further afield – as being the bird of doom. This was well known in the 17th century in Shakespeare's time. In his play *Macbeth*, Lady Macbeth declares that the Raven 'croaks the fatal entrance of Duncan', Duncan the king, who is later murdered by Macbeth. It is not so well known that the Raven was felt to have an ambivalent character, sometimes good, sometimes evil. Scholars believe this ambivalence can be traced historically to two traditions, Judaism and Christianity (the good), and heathen (the bad). Whichever way people saw the Raven, they believed it was a supernatural bird. So, when it flew 'back and forth' on its release, it appears on our first reading of this that it did not return to Noah in the ark and so it perished at sea. However, as a supernatural creature Noah would have believed it had miraculously survived, and that augured well for him.

Kites and vultures must have been intended in this prophetic image:

 HABAKKUK
CH. 1, VV. 8-9

They fly like a vulture swooping to devour;
they all come bent on violence.
fiercer than wolves at dusk.
Their cavalry gallops headlong;
their horsemen come from afar.

What is unusually interesting is not that the prophet repeats a description of the Babylonians' proverbial speed of attack, but that the verses are mentioned in the *Commentary on Habakkuk,* which was among the first group of scrolls found accidentally near Qumran in 1947, and now known as the *Dead Sea Scrolls*. Qumran was a monastic style settlement of the Essenes, a Jewish sect. Carbon-14 and palaeographic tests on the scroll point to this copy having been written in the first century BC. The writer says of these verses:

 DEAD
SEA SCROLLS
A NEW TRANSLATION BY,
WISE, ABEGG JR, AND COOK

[This refers to] *the Kittim who trample the land with [their]*
horses and their beasts. From far away they come, from the
seacoasts, to eat up all the peoples like an insatiable vulture.

Scholars agree that the commentator here is referring to the Romans (i.e. the Kittim) who will come from overseas. They did, in 63 BC, and several years later they destroyed Qumran. About 500 years after Habbakuk had written, a member of the Essenes still thought the image of devouring vultures

Raven

was a perfect description of the enemy. He added that he – and no doubt his contemporaries too – considered that vultures were 'insatiable'. Griffon Vultures are so large that their greedy feeding habits as they crowd around food are particularly noticeable. The word is a welcome addition to the otherwise regular comment, and is probably an observed detail of the vulture's behaviour, which the ancient authors had not described.

The eagle is so deeply fixed in the religion, mythology and ritual of many lands, especially Europe, North America and the Middle East, that even today many people automatically think of the eagle as the king of birds, and that any large bird of prey is an eagle. We still speak of someone being 'eagle-eyed' to describe how observant they are. That is a faint echo of the sympathetic magic of ancient times, which enabled the power of the creature to enter the person.

Although these birds featured so prominently in the lives of the Jews, the Bible record is sometimes unclear as to whether an eagle or a vulture is intended, because the word 'nesher' is used for both. This is not so surprising because these huge birds are superficially alike, especially in the air high above the observer. A careful consideration of what the bird called 'nesher' is doing does, however, often make it clear to the reader that the bird's behaviour is certainly that of either an eagle or a vulture. Generally speaking, the eagle will be the one hunting live prey and the vulture (or more likely, vultures) will be feeding on something already dead. There is no confusion today, because the modern Hebrew says 'nesher' for the Griffon Vulture, and 'ayit (plus a descriptive word)' for each of the eagles.

The Hebrews were part of a Middle Eastern culture which, together with their neighbours, the Babylonians, the Assyrians and the Hittites, held eagles in high esteem. Later, the Roman, scientific writer, Pliny the Elder (23–79 AD), described how many of these older nations felt about these birds: 'Eagles carry the price for both honour and strength'. Edward A. Armstrong, the author of *The Folklore of Birds*, regards the Middle East as the centre from which eagle traditions radiated. We call an eagle the king of birds, perhaps without understanding why. The Hebrews did know why. Eagles were symbolic of power and they wished that the power would brush off on them, and so enable them to win in their struggles with their enemies. Unfortunately for them, that did not always work out as they wished.

A psalm of David describes how the Jews would think about an eagle, the men especially:

> *Praise the Lord, O my soul;*
> *all my inmost being, praise his holy name.*
>
> PSALM *Praise the Lord, O my soul,*
> 103, vv. 1-5 *and forget not all his benefits ...*
> *who satisfies your desires with good things*
> *so that your youth is renewed like the eagle's.*

Golden Eagle

All these benefits – forgiveness, good health, redemption, love, compassion, and feeling good – will restore a person's youthful vigour and match it to what was believed to be the unfailing strength of the eagle. The writer of the second half of Isaiah expressed the same idea many years later:

≈ ISAIAH
CH. 40, VV. 29-31

[God] gives strength to the weary
and increases the power of the weak.
Even youths grow tired and weary,
and young men stumble and fall;
but those who hope in the Lord
will renew their strength.
They will soar on wings like eagles;
they will run and not grow weary,
they will walk and not be faint. ≈

TAWNY EAGLE

The characteristic of an eagle which the biblical writers repeatedly refer to, and which no doubt led to its being thought of as the source of great strength, was its flight. The writer of Proverbs summed up the wonder that he and many others felt. Of the four things in the world he did not understand, the first was 'the way of an eagle in the sky' (Proverbs ch. 30, v. 9). Whether the 'way' is where the eagle is going, for there are no tracks in the sky which show its route, or whether it is the manner in which it is managing to stay aloft so effortlessly, it is a sight which is hard to understand. It is indeed an awesome sight to see an eagle drifting, or circling on motionless wings high above the observer, or diving to capture prey. To be able to fly like an eagle was greatly to be admired and wished for, and its speed and strength were the two characteristics the Jews admired most.

SPARROW-
HAWK

The writers of the Old Testament who watched eagles must have thought like that. The Jews' deeply held beliefs about the eagle's power go back to their understanding of creation. The first chapter of Genesis describes how the 'Spirit of God was hovering over the waters' (Genesis ch. 1, v. 2). This awesome and majestic picture of the way God is caring for His creation was clearly in Moses' mind when he recited his great song of praise.

The time when the Hebrews of old would have been most likely to see 'eagles swooping down on their prey' (Job ch. 9, v. 26) was in autumn when large numbers pass through Palestine on their way to winter in East Africa. Two species in particular, in modern times anyway, are especially common. There is no reason to believe that times have changed. The eagles of the genus *Aquila* are among the largest, the Golden Eagle *Aquila chrysaetos* included. Firstly, birdwatchers counted nearly 142,000 Lesser Spotted Eagles *A. pomarina* passing south at Kefar Kassem between the end of August and mid-October in 1983. This eagle has a wingspan of up to 168 cm (70 in). Even larger is the Tawny or Steppe Eagle *A. rapax*, whose broad, dark brown wings span up to 190 cm (79 in) and are a dramatic sight in the sky. It, too, is seen in large numbers on migration. In southern Israel at Eilat around 29,000 were identified passing north in the spring of 1977. With so many of them drifting by on thermals of warm air, or hunting by the wayside, it is no wonder that the writer of Proverbs, maybe Solomon himself, wrote:

PROVERBS   *for they will surely sprout wings*
CH. 23, v. 5   *and fly off to the sky like an eagle.*

Common Buzzard

One of the most spectacular aspects of migration through the Holy Land is the vast number of birds of prey, not just eagles, which pass through each spring and autumn. Falcons and hawks are all involved too. Two species of hawks in particular pass through, the Common Buzzard *Buteo buteo* and the Levant Sparrowhawk *Accipiter brevipes*. The former in modern times is the commonest migrating raptor in spring; more than 100,000 were counted passing over Eilat on one day in March 1980; it is not so common in autumn. But the Levant Sparrowhawks migrate in autumn – going south in flocks of hundreds or even thousands. A very obvious reference to migration is in one of the questions that God asks Job:

JOB
CH. 39, V. 26

Does the hawk take flight by your wisdom
and spread his wings towards the south?

The climax is in the third week of September when between 1,000 and 3,000 have been counted in a day at Kafar Kasem on the western slopes of the central mountain range. This route was not discovered by modern ornithologists until 1977, but it seems clear that the writer of Job knew all about it. A Levant Sparrowhawk seems to be the most likely species for the hawk named in verse 26. An individual ringed by scientists in Eilat was found two years later in Romania. Although we have seen that the naming of birds by the ancient writers has created problems for modern translators, the identification of this bird as a 'hawk' is secure. The Bible has the word 'nets'; in modern Hebrew the hawks are all 'nets' (or 'nez' as a modern bird list spells it), plus a qualifying adjective.

Perhaps the most splendid of all the mentions of an eagle is from Proverbs, where the writer declares:

BEARDED
VULTURE

PROVERBS
CH. 30, VV. 18-19

There are three things that are too amazing for me,
four that I do not understand:
the way of an eagle in the sky,
Cast but a glance at riches, and they are gone,
the way of a snake on a rock,
the way of a ship on the high seas,
and the way of a man with a maiden.

I am sure that eagles were a rich part of the Hebrews' daily experience.

But what is the 'ossifrage' which is listed in the *Authorised Version* translation of the Leviticus list of birds? The name is formed by the adaptation of two Latin words meaning 'bone-breaker'. And that is exactly what this bird does. Today we know it as the Lammergeier or Bearded Vulture *Gypaetus barbatus*. It is a bird of rocky, mountainous regions. Its old name comes from its habit of picking up a bone from a carcass, flying up to 80 m (250 feet) or so, and dropping the bone onto rocks to be

PEREGRINE FALCON

smashed, so enabling the bird to get at the marrow and small bits of bone, which constitute 70 per cent of its diet. Although listed in the *Authorised Version* I can find no specific reference in the Bible to its amazing bone-breaking technique. It is a very rare resident now in the Judean Desert and the Negev. It has decreased in modern times, as have the other vultures, yet the Rev. Tristram recorded its breeding in the cliffs near the Sea of Galilee, and 'most of the ravines are peopled by a pair'. Modern translations do translate 'peres' as 'vulture' but not specifically as the Bearded Vulture as Dr Driver or the *Checklist* do. Its appearance, rarity in Europe and its habit, make it a bird which birdwatchers today particularly wish to see. The Bearded Vulture is still known as 'peres' in Israel today.

The listing of 'kite' gives us another bird which feeds on carrion. The attractively coloured Red Kite *Milvus milvus* was formerly common according to Rev. Tristram but is now rare and seen only on passage. The Black Kite *M. migrans* was quite a common breeding resident until the 1960s, but now only a few pairs can be found in the north. It is common on passage in spring and autumn; it is widespread in Europe (not Britain or Scandinavia) and Asia. This is the bird which gathers in large numbers at rubbish tips in the East. Although it is not referred to outside the list in Leviticus, we can understand from its scavenging habits why the kites would have been considered 'unclean'. Its identity seems sure – the Hebrew word used is 'daah' or 'dayyah', and the modern Hebrew is 'daya mezuya' for the Black Kite (other kites are 'daya' plus an appropriate descriptor).

The other hawks, buzzards and falcons are only rarely referred to in the scriptures. Job in chapter 28 writes a poem, trying to answer the question 'Where can wisdom be found?' He makes several suggestions and then he writes:

No bird of prey knows that hidden path,
no falcon's eye has seen it.

Red Kite

Falcons have wonderful eyesight. Wisdom is so difficult to find that even a falcon's eyesight cannot find it. There are 11 species of falcon on the Israeli list. The commonest is the Kestrel *Falco tinnunculus*, which is a familiar sight in Britain too, hovering by a roadside looking for a vole or a large beetle. Perhaps the most impressive falcon to watch is the Peregrine *F. peregrinus*, specialist hunter of pigeons. As it dives to make a kill it has been timed at well over 160 km/h (100 mph), making it one of the fastest, if not the fastest, bird on Earth. It has decreased in modern times in Israel, as it has in many countries, because of pesticides. It used to nest on Mount Carmel until the 1950s. Now it is quite common on passage. I like to think that it was this falcon that so impressed Job.

The Osprey *Pandion haliaetus* is a surprise inclusion in the Leviticus list, if it is correctly identified. It is almost completely a fish-eater, spectacularly diving to capture a fish with its outstretched talons. Yet the Law clearly stated that 'of all the creatures living in the water of the seas and the streams, you may eat any that have fins and scales'. Fish are not unclean. So the Osprey seems to be listed with the other flesh-eaters simply because it does eat blood, the life of the fish. Only a few nest in Israel, and it is scarce on passage now.

The final reference to these raptors is in the last book of the Bible. The vision of John the Divine was written when Christians were suffering persecution from the Romans towards the end of the first century AD. He sees the Four Horsemen of the Apocalypse, who were first described by the prophet Zechariah. When John sees the White Horse and its rider, Faithful and True, the symbol of conquest, an angel appears too, who:

REVELATION
CH. 19, VV. 17-21

cried in a loud voice to all the birds flying in mid-air,
"Come, gather together for the great supper of God,
so that you may eat the flesh of kings, generals and
mighty men, of horses and their riders, and the flesh
of all people, free and slave, small and great." ...
and all the birds gorged themselves on their flesh.

So 'nature's ghastly gourmet' features in the deadly military campaigns which involved the Israelites in struggles against Egyptians, Assyrians, Babylonians and Romans. Against these nations and vultures, the eagle had no reply until followers fully understood Christ's mission, thanks to the teaching and support of such people as the writers of the Gospels, Paul and Silas, Lydia the seller of purple, and John.

The writer of Revelation well understood the Roman's cruel persecution because he wrote the book in wretched exile on the island of Patmos, writing in the 'code' of apocalyptic literature familiar to the Jews (such as the book of Daniel), but whose true meaning would not have been understood by the Romans. John has a vision of the Throne of Heaven around which are:

Black Kite

Male (left) and Female lesser Kestrels, Summer Visitors to Israel

REVELATION
CH. 4, VV. 6-7

four living creatures, and they were covered with eyes, in front and behind. The first living creature was like a lion, the second was like an ox, the third had a face like a man, the fourth was like a flying eagle.

Since then, as the Christian church became more established, the word of God has been spoken by readers in church at the lectern, which holds the Bible. The top of that lectern is often a carved wood or moulded brass eagle with outstretched wings. The writers of the four Gospels have for many centuries been depicted in art with a symbol: Matthew, a winged man; Mark, a winged lion; Luke, a winged ox; and John – the eagle.

Leviticus also mentions herons as unclean birds, presumably because they spilt the life blood of their prey, although a favourite food of them all was fish, which was not unclean. The heron family includes the egrets. They nest in colonies and several are common on passage in spring and autumn, and others are fairly common or common residents. The commonest now is the Little Egret *Egretta garzetta*, which is abundant in several areas. It is easily seen; it feeds in cultivated regions, where it preys on 'unclean' insects, frogs and small fish. Next most common is the Cattle Egret *Bubulcus ibis*,

CATTLE EGRET

SACRED IBIS

which now breeds, although it used to be a common winter visitor before that. They feed on 'unclean' creatures too, disturbed by the farm animals, as its name implies. It did not nest in Israel until the early 1950s. The Squacco Heron *Ardeola ralloides* is another small member of the heron family; it is mainly a common passage migrant through Israel, with hundreds passing through in March–April and September especially, stopping to feed in the wetlands. It was not recorded breeding in Israel until 1959.

The Bible does not have many records of birds that *were* eaten. The wisdom and riches of Solomon were widely known, as the visit of the Queen of Sheba testifies. What is perhaps not so well known is the fact that Solomon used much of his wealth to feed all his household, his servants, court officials and their families. The daily provisioning of all these people was a huge task for his district officers. They had to be sure of having at hand, every day

thirty cors [about 5,000 litres/1,100 gallons] *of fine flour and sixty cors* [about 10,000 litres/2,200 gallons] *of meal, ten head of stall-fed cattle, twenty of pasture-fed cattle and a hundred sheep and goats, as well as deer, gazelles, roebucks and choice fowl.*

✒ 1 KINGS
CH. 4, VV. 22-23

It is only natural to think that having been given the list in Leviticus of birds not to eat, we would be able to find a list of birds which are good for the table, and it would be straightforward to read somewhere exactly what species of 'choice fowl' Solomon ate. But that is not so. To learn what birds might have been on the menu in Biblical times is even more of a detective story than the investigation regarding birds' identities because of the difficulties in translation, as we have seen.

Historians and scientists are reasonably sure that our 'chicken' is descended from the Red Jungle Fowl *Gallus gallus* of the forests of Southeast Asia. Early civilization in the Indus valley of India had certainly domesticated them by about 2500 BC. By 1500 BC it is believed that this valuable bird had spread north into China, and as far west as Egypt in tribute paid by Babylon to Pharaoh Thutmose the Third. Solomon reigned in Palestine from 970–930 BC. So, what we know as the 'chicken', could have been on Solomon's table because when the Queen of Sheba saw 'the food on his table ... she was overwhelmed' (1 Kings ch. 10, v.5), which suggests his menu was sumptuous and rare. It is well known from written evidence and wall paintings that the Egyptians trapped geese and fattened them.

WHITE-FRONTED, RED-BREASTED AND GREY-LAG GEESE FROM EGYPTIAN WALL PAINTING C. 2600 BC

Solomon may have copied their example. In some commentaries is what seems to me to be a rather curious interpretation of the Hebrew word 'barburim' used here for 'fowl'; they say that the 'choice fowls' were Cuckoos *Cuculus canorus*, whose characteristic call is said to be imitated in the Hebrew word. Cuckoos were an ancient delicacy. They are migrants through Palestine, from Europe and south-west Asia, on their way to and from southern Africa. But Solomon's district officers could not possibly have provided them as part of the king's *daily* provisions. A more realistic alternative would be that the birds were guinea fowl. They are widespread today in the savannahs of Africa south of the Sahara. Their domestication was well known in ancient Greece, and the Romans carried on the practice; a Roman mosaic in Cyprus clearly shows a Helmeted Guinea Fowl *Numida meleagris*, the same, white-spotted, grey bird we sometimes see today noisily wandering around a farmyard. The species probably had a more northerly distribution in ancient times when the Sahara was not so widespread and could well have become centuries earlier a favoured dish of the rich, even in Israel.

The Israelites did trap birds, almost certainly ground-feeding species such as pigeons and doves, partridges, and Quails. There are several references to nets and snares being used, admittedly here in these quotations, mostly to illustrate how easily a sinful or careless person could be caught. But all the writers, not just the writer of the first example which is a statement of common sense, must also have been writing from experience of the real thing for them to have created such vivid images:

PROVERBS
CH. 1, V. 17

How useless to spread a net
in full view of the birds!

PROVERBS
CH. 7, V. 23

All at once he followed her...
like a bird darting into a snare

JEREMIAH
CH. 5, V. 26

for among my people there are wicked men
who lay snares like a fowler's net

JEREMIAH
CH. 3, V. 5

Does a bird fall into a trap on the ground
if the striker is not set for it?

The last one seems to refer to what we call today a 'clap net'. It is used still by birdcatchers, and ornithologists who are interested in migration studies and need to fit a small identity ring on the birds' legs. This net was popular in Egypt, and so probably in Palestine too. It would be laid on the ground, covering an area of 8 sq. m (80 sq. ft) or more, and would be closed through 180 degrees at a given signal by pulling on a rope. All three kinds of birds mentioned above are attracted to seed, so maybe a food source was naturally or artificially present and would be an ideal catching site.

In Israel the common, wild pigeons likely to be in flocks in Biblical times are the Rock Dove

Rock Dove

Columba livia (the ancestor of our feral town pigeons), Stock Dove *C. oenas*, Wood Pigeon *C. palumbus*, and Turtle Dove *Streptopelia turtur*. In Rev. Tristram's time, and no doubt centuries before, the gregarious Rock Dove could be seen in thousands in Jericho, in hundreds near Hebron, and 'words cannot convey an impression of the multitude' in the region of Nahal Amud. Today it is much

STOCK DOVE

less common. The Stock Dove and Wood Pigeon are mostly winter visitors to Palestine, a few in some years and many in other years, sometimes in flocks of hundreds – just what the Israelite fowlers would have prayed for. Two other doves are common in Israel today, the Collared Dove *Streptopelia decaocto* and the Palm Dove *S. senegalensis*. Although these two small, confiding doves of the towns are commonly seen today by visitors and pilgrims, they would not have been hunted by the Israelites because they arrived in Israel after Bible times. It is a very different story with regard to the migratory Turtle Dove, which is, and so far as we know always has been, a common summer visitor – in three minutes I have seen around 200 in April on telegraph wires by the road in the Jordan valley, and have heard many from Jerusalem to Caesarea Philippi repeating their purring song. The doves and pigeons would have been good to catch and eat. Only the Turtle Dove is specifically named.

Four members of the pheasant family are still popular game-birds and would certainly have attracted the attention of Palestinian fowlers and egg collectors; they are the Chukar and Sand Partridge *Alectoris chukar* and *Ammoperdix heyi*, the Black Francolin *Francolinus francolinus* and the Quail *Coturnix coturnix*. The first is the most common today in fruit groves, fields and open ground. The second is locally common in the Jordan valley in rocky places south of the Sea of Galilee as far as the Negev. The Francolin is a bird of dense vegetation, farmed or wild, in the Jordan valley. The migratory Quail passes through in spring and autumn in great numbers, and many stay to breed. It is a small bird, only 16–18 cm (6–7 in) long, and is easily hidden in the pasture or wild grassland which it favours, but its three note call gives it away – '*wet-my-lips, wet-my-lips*'. There are two stories in the Old Testament that retell the miraculous arrival of many Quails, just when the Israelites were moaning to Moses, "If only we had meat to eat!" The first is in Exodus chapter 16, and the second in Numbers chapter 11. This may seem an extraordinary, even impossible, occurrence. However, the Quail is the only migrant member of the pheasant family which breeds in Europe and the Middle East; it migrates south to spend the winter south of the Sahara, and even today, in some years, it is extremely abundant along the Mediterranean coast. In times past it has been known to arrive in some areas in flocks of several thousands. It is still a much sought-after bird for food, and flocks now are more likely to be numbered in the tens. As recently as between 1920 and 1930, during the autumn migration, the Egyptians are known to have trapped up to a million annually. The Israelites in The Wilderness certainly did enjoy a huge feast of Quails!

There is a very revealing hunting illustration to be found in a conversation that young David had with King Saul. David was a fugitive from Saul's jealous anger, and

WOOD PIGEON, IMMATURE BEHIND ADULT

had the chance to kill him but refused to, for as he said, "Who can lay a hand on the Lord's anointed and be guiltless?" (1 Samuel ch. 26, v. 9). David shouted to Saul's guard and Saul heard him and called out to him. David asked why Saul was chasing him, and of what was he guilty. He ended by saying, "The king of Israel has come out to look for a flea – as one hunts a partridge in the mountain" (1 Samuel ch. 26, v. 20). David was suggesting that the king was making a fool of himself in his

fanatical pursuit of an innocent man. His figure of speech suggests a big hunting party out to catch a very small item of prey. The most likely partridge that David had in mind is the Sand Partridge, which is resident in the wild country in the Dead Sea depression that David was in. The species is still called 'qore' in Hebrew, the very word used in the Old Testament stories. Both species are still hunted for food and sport.

The prophet Jeremiah is the only other writer who names the partridge, in a rather strange proverb:

> JEREMIAH *Like a partridge that hatches eggs it did not lay*
> CH. 17, V. 11 *is the man who gains riches by unjust means.*

Partridges are actually very good parents, unlike the closely related Pheasant *Phasianus colchicus*, and do not steal other eggs, or lay eggs in others' nests as the Cuckoos. The proverb is based on an ancient belief that a hen partridge did steal eggs and put them in her own nest – how else could one account for such a large clutch? Ornithologists today have studied partridges and know that a Chukar

SAND PARTRIDGE

will commonly lay 8–15 eggs; up to 20 is not unknown, and the Rev. Tristram discovered a nest with 26. The Chukar is common to very common throughout all Israel today; it is found in fields, orchards, groves and all kinds of open ground. It is a very good candidate for the bird to which Jeremiah was referring. Other references do not name a species but do suggest that the Israelites well knew game birds' nests and took their eggs for food. Through the words of the prophet, God describes how He will defeat Assyria:

I plundered their treasures …
As one reaches into a nest
So my hand reached for the wealth of nations,
as men gather abandoned eggs,
so I gathered all the countries;
not one flapped a wing,
or opened its mouth to chirp.

ISAIAH
CH. 10, vv. 13-14

I do not find it so surprising that there are few references to bird *song* in the Bible, for, as Alice Parmelee remarked in her definitive book *All the Birds of the Bible* in 1959: 'Bird song is not heard by many people, even though it is there, because they do not 'tune in' to it.' In spring there is plenty of birdsong in Palestine, in the fields, in the hills, by the River Jordan, and even in the villages and towns. In just the first two days of a stay in Jerusalem in mid-April I heard Palm Doves cooing, Swifts screaming in their courtship flights, several Yellow-vented (or Spectacled) Bulbuls singing, their powerful, rich, flutey notes ringing across the roofs of Jerusalem, a Great Tit was singing '*teacher-teacher*' in the Garden of Gethsemane, and in the leafy sanctuary of the Garden Tomb, a Blackbird, a Greenfinch and Palestine Sunbird were singing – the last in a flowering Judas Tree. So, I think there would have been plenty of birdsong to listen to in the Psalmist's and Christ's time – if they had been as aware of that as their own music and song which the Psalmist writes about in Psalms 98 and 150.

Sadly there is only one other reference to birdsong in the Bible, which we read of earlier in a slightly different translation:

For now the winter is past,
the rains are over and gone;
the flowers appear in the countryside;
the time is coming when the birds will sing,
and the turtle dove's cooing will be
heard in our land.

SONG OF SONGS
CH. 2, vv. 11-12 (NEB)

With these words the Beloved appeals to her Lover. He replies with another, gentle image of a dove:

My dove in the clefts of the rock,
in the hiding places on the mountainside,
SONG OF SONGS *show me your face,*
CH. 2, V. 14 *let me hear your voice;*
for your voice is sweet,
and your face is lovely.

The lovers really were well aware of, firstly, what was surely the soothing, purring song of a Turtle Dove, and then of the 'sweet' cooing of the Rock Dove.

In a hymn in praise of and longing for the Temple by a writer who was denied access to the Temple during the ravaging of Judah by Sennacherib in 701 BC (described in 2 Kings ch. 18), we read this:

How lovely is your dwelling place
O Lord Almighty!
My soul yearns, even faints,
for the courts of the lord;
PSALM *my heart and flesh cry out for the living God.*
84, vv. 1-4 *Even the sparrow has found a home,*
and the swallow a nest for herself.
where she may have her young –
a place near your altar,
O Lord almighty, my King and my God

BARN SWALLOW

Even though the writer, probably a Levite who would have normally officiated in the Temple, is distressed because he cannot go into the Temple, he is near enough and observant enough to see the birds which do gain access, or he is remembering what he has seen in the past when he was able to go into the Temple – and the birds' finding a 'home' emphasizes his distress. Both the House Sparrow *Passer domesticus* and the Barn Swallow (the one we usually call simply 'Swallow') *Hirundo rustica* do nest in buildings; House Sparrows are often found in a loose colony, in holes in a wall or under roof tiles. The wild site for a Swallow would have been a cave entrance, where it would stick its mud nest to a ledge, but since time immemorial Swallows have nested in man-made structures such as barns, bridges, church porches and sheds – or a Temple. The Psalmist may have seen either Barn Swallow or Red-rumped Swallow *H. daurica*, or both, because both commonly nest in buildings and are still widespread in summer in the Holy Land.

In Jeremiah chapter 28, he lists four birds. The fourth Hebrew name is 'sus'. This has been translated in times past as 'swallow'. But Jeremiah is quite clear that his birds are migrants, and would be more noticeable than Swallows, which do winter in Israel. The Rev. Tristram discovered that local Arab boys identified the hundreds of Swifts *Apus apus*, which were swarming about Mount Carmel and the surrounding towns and villages as 'sis'. Accompanied by screaming calls, the dramatic aerial chases of courting Swifts is a spectacular sight. Their 'appointed time' would be clearly signalled by these black, dashing birds. Today the four species of swift that have been recorded in Israel are still known as 'sis' (plus a different descriptive word for each species). The Common Swift breeds in small colonies under the eaves of houses, in holes in walls, or gaps between shutters, at a safe distance above the ground. The Pallid Swift *A. unicolor* is a less numerous summer visitor to Israel. It nests only in caves and rocky clefts, in the Negev and Golan heights for example. Anyone today who visits Jerusalem or Nazareth or Tel Aviv in spring cannot fail to notice these dark, exciting birds, sometimes both species together in migrating flocks. Jeremiah must have observed them too.

Our last mention of a calling or singing bird is an unusual one. Among God's great list of questions to Job is a series about the huge marine animal that He calls Leviathan, and one of the questions asks is:

RED-RUMPED SWALLOW

This is the only reference in the scriptures to cage birds, but it is a fascinating insight into what was probably daily life in Palestine. Although the Bible does not say much about the joy people can feel from hearing birdsong, this one question implies that song birds in cages were a familiar sight in ancient times. Other ancient historical records mention them. King Hezekiah of Judah in the seventh century BC was besieged in Jerusalem 'like a caged bird'. Other people who lived by the Mediterranean also kept cage birds as is shown on classical Greek vases, which are 2,500 years old. Anyone who recently has had a holiday there will have seen little cages hanging on the walls of the houses, with maybe a Canary *Serinus canaria* as we would have at home, but often a Goldfinch *Carduelis carduelis* or a Bullfinch *Pyrrhula pyrrhula*. The Yellow-vented Bulbuls of Palestine are not beautifully coloured but they are handsomely marked, their white eye-rings bright against the black head, and the bright yellow under the tail in great contrast to the grey-brown plumage. They are talkative birds! Their natural, flute like song is enhanced by a wide range of calls, which mimic other species. They are a noticeable feature of Palestinian birdsong, as we have already read, and could well be the species God had in mind.

Several times in this book Rev. Tristram has been mentioned, and we will close with a memorial to this 19th-century worthy who was among the first to study and write at length about the wildlife of the Holy Land. Today he is remembered in the region by two birds in particular. Tristram's Serin *Serinus syriacus* is a close relative of the Canary, and has a song to match. It breeds commonly on Mount Hermon where traditionally Jesus was Transfigured (see St Matthew ch. 17, St Mark ch. 9, St Luke ch. 9). The other bird is Tristram's Grackle *Onychognathus tristrami*, which the Jews today call 'Tristramit'. It is a relative of the familiar Starling *Sturnus vulgaris*. It has a limited distribution in the Arabian Peninsula, and in more recent times has spread north to near Jericho. Pilgrims today to the Holy land may observe it closely, as I have, at the fortress of Masada overlooking the Dead Sea. The Essenes, who were a Jewish religious sect, had a monastic-style settlement at Qumran to the north of Masada. Some scholars say that John the Baptist and Jesus may have had links with them. So, who knows, maybe the cousins saw these distinctive, red-winged, black birds.

CHAPTER VI
OTHER CREATURES

In the beginning of the world, we read in Genesis, the Bible's first book, that God created birds and animals of every kind. The snake is actually the first creature named in The Bible. The English language is rich in synonyms – witness this list of words, which, on the face of it, all describe the same creature – snake, serpent, adder, viper, cobra and asp. Then when we come to read about these reptiles in the Bible we discover that at least nine Hebrew words for this animal are listed in Strong's *Exhaustive Concordance*, which compound the problem for translators in their search for the correct English name. The roots of the Hebrew words describe the animal's behaviour, so the context of the word helps translators find the right species. One could be forgiven for thinking that Satan, the Devil, had got amongst the text to upset things just as he did in the Garden of Eden, in the form of a serpent (or snake or adder, even different editions of *NIV* do not agree!):

GENESIS
CH. 3, VV. 1-5

Now the snake was more crafty than any of the
wild animals the Lord God had made.
He said to the woman, "Did God really say,
'You must not eat from any tree in the garden'?"
The woman said to the snake, "We may eat fruit
from the trees in the garden, but God did say,
'You must not eat fruit from the tree that is in the
middle of the garden, and you must not touch it,
or you will die.'"
"You will not certainly die," the snake said to the woman.
"For God knows that when you eat
from it your eyes will be opened, and
you will be like God, knowing good and evil."

The woman is persuaded by the snake to eat the forbidden fruit and she also gives it to Adam. Their eyes are opened to good and evil, and when God finds out about this He says to the woman:

"What is this you have done?"
The woman said, "The snake deceived me, and I ate."
So the Lord God said to the snake, "Because you have done
this, cursed are you above all livestock and all wild animals!

COBRA (BELOW)

*You will crawl on your belly and you will eat dust all the
days of your life. And I will put enmity between you and the
woman, and between your offspring and hers; he will crush
your head, and you will strike his heel."
To the woman he said, "I will make your pains in
childbearing very severe; with painful labour you will give
birth to children. Your desire will be for your husband,
and he will rule over you."
To Adam he said, "Because you listened to your wife and ate
fruit from the tree about which I commanded you, 'You must
not eat from it', cursed is the ground because of you; through
painful toil you will eat food from it all the days of your life.
It will produce thorns and thistles for you, and you will eat
the plants of the field. By the sweat of your brow you will eat
your food until you return to the ground, since from it you
were taken; for dust you are and to dust you will return."
Adam named his wife Eve, because she would become the
mother of all the living.*

GENESIS
CH. 3, VV. 13-20

The snake (serpent) is a symbolic creature in many cultures across the world, from Ancient Egyptian to Norse, Greek, Indian and Australian mythology. Adam and Eve's meeting with it has resulted in the Jewish and Christian faiths seeing the episode as a way of describing very difficult concepts which we still struggle to understand today. How do you and I describe mankind's fall from God's grace, the power of temptation and the source of that power, and the difference between good and evil? In the Genesis story God's different punishments for Adam, Eve and the snake are all ways of explaining basic questions of life and death – Adam will have to struggle to find food to live all his life, childbirth for Eve will be painful, the struggle between man and the snake is the ongoing fight between good and evil, and the snake will always be cursed and a sign of evil – particularly as mankind got to know the deadly nature of poisonous species and those which squeeze their prey to death. It even has to grovel to God by crawling on its belly. Still today in English we describe a false friend or a lurking danger as 'a snake in the grass', which is based on an ancient Latin proverb.

The Israelites complained to Moses about the lack of food in the wilderness (p. 28) and God sent them manna. There was another miracle too:

*The people grew impatient on the way, they spoke against
God and against Moses, and said, " Why have you brought us
up out of Egypt to die in the desert? There is no bread! There*

NUMBERS
CH. 21, vv. 4-9

is no water! And we detest this miserable food!"
Then the Lord sent venomous snakes among them; they bit
the people and many Israelites died. The people came to
Moses and said, "We sinned when we spoke against the Lord
and against you. Pray that the Lord will take the snakes away
from us." So Moses prayed for the people. The Lord said to
Moses, "Make a snake and put it on a pole; anyone who is
bitten can look on it and live." So Moses made a bronze snake
and put it up on a pole. Then anyone who was bitten by a
snake and looked at the bronze snake, he lived.

St John in his Gospel reminds his readers of this event when he retells Jesus' teaching to Nicodemus about being born again. The first hearers of John's gospel would have been well aware of the new life that God gave the Israelites in the desert, and may well have been startled by Jesus stating that He was going to be lifted up like the snake, but on a cross as John recorded, and believers in Christ's message would be 'born again', and they 'may have life, and have it to the full' (St John, ch. 10, v. 10).

Moses features in two other snake stories, in Genesis, chh. 4 & 7. Elsewhere the widely believed evil nature of the snake is sharply used in a description of unjust rulers:

PSALM
58, vv. 1-5

Do you rulers indeed speak justly?
Do you judge uprightly among men?
No, in your heart you devise injustice,
and your hands mete out violence on the earth.
Even from birth the wicked go astray;
from the womb they are wayward and speak lies.
Their venom is like the venom of a snake,
like that of a cobra that has stopped its ears,
that will not heed the tune of the charmer,
however skilful the chanter may be.

Sadly, all too often we hear in the news today the same cry of anguish in many parts of the world, for example from persecuted Christians in several countries in Africa and Asia.

Snakes generally are linked to descriptions of mortal threats:

> *I will say to the Lord, "He is my refuge and my fortress,*
> *my God, in whom I trust ..."*

PSALM
91, vv. 2-13

> *You will not fear the terror of night,*
> *nor the arrow that flies by day ...*
> *You will tread upon the lion and the cobra;*
> *you will trample the great lion and the serpent.*

The Psalmist carefully balances the emphatic list of dangers with the power of God's help, which will positively aid the believer. There are two Hebrew words in the original, which have taxed translators ever since, hence the two English words 'cobra' and 'serpent'. Our quotation above is from *NIV*; the *Authorised Version* of 1611 finishes verse 13 with '... the lion and the adder' and 'the young lion and the dragon', and *The Good News Bible* speaks of 'lions and snakes' and 'fierce lions and poisonous snakes'. If 'cobra' is right, it could be the Egyptian or Hooded Cobra *Naja haje*, which is found across North Africa and the Israelites would have got to know it in Egypt, but Rev. Tristram recorded it as rare in Palestine. He does list over two dozen species of snake in the region, including several species of Viper *Vipera* spp. That name is used to describe snakes in several stories, perhaps most dramatically here when St Paul was shipwrecked on his way to Rome:

ACTS OF THE APOSTLES
CH. 28, vv. 1-5

> *Once safely on shore, we found out that the island was*
> *called Malta. The islanders showed us unusual kindness.*
> *They built a fire and welcomed us all because it was raining*
> *and cold. Paul gathered a pile of brushwood and, as he put*
> *it on the fire, a viper, driven out by the heat, fastened itself*
> *on his hand. When the islanders saw the snake hanging*
> *from his hand, they said to each other, "This man must be*
> *a murderer; for though he escaped from the sea, Justice*
> *has not allowed him to live." But Paul shook the snake*
> *off into the fire and suffered no ill effects.*

The islanders then thought he was a god! The snake's identification as a viper goes back to *The Authorised Version*, but *The Good News Bible* uses the more general term 'snake'.

St Paul was convinced that all were sinners and needed faith in Jesus Christ to become righteous. He described people's sinfulness as:

ROMANS
CH. 3, v. 13

> *Their throats are open graves;*
> *their tongues practise deceit.*
> *The poison of vipers is on their lips.*

The *Authorised Version* and *The New King James Bible* speak of 'asps' not vipers and *Good News* refers to 'snake's poison'. Asp is the anglicised word from Latin 'aspis' which in ancient Egyptian and Roman times was used to describe any of several poisonous snakes. Famously it became the killer of Cleopatra of Egypt. There is a reference to England's only poisonous snake, the Adder, in Jacob's prophetic blessing to his sons:

GENESIS 4
CH. 9, V. 17
(AUTHORISED VERSION)

Dan will be a serpent by the way, an adder on the path, that biteth the horse's heels, so that his rider shall fall backwards.

HORNED VIPER

The treachery of the tribe of Dan is told later in Judges chapter 18. *NIV* says 'serpent' and 'viper' and *Good News* once again vaguely talks about 'snake' and 'poisonous snake'. In English 'adder' and 'viper' are synonymous, and the 17th century translators can be forgiven for using 'adder' because they would not have had experience of other poisonous snakes, and their readers – and many readers today – would certainly understand a poisonous snake was meant. This reference has even been thought to mean the Horned Viper *Cerastes cerastes*, which is widespread across the deserts of North Africa and the Middle East. It is easily recognised by the horn over each eye and its 'side winding' movement.

After so many references to the disagreeable nature of snakes, the following statement may seem to be a surprise, coming as it does in Jesus' instructions to his disciples:

ST MATTHEW
CH. 10, V. 16

I am sending you out like sheep among wolves. Therefore be as shrewd as snakes and as innocent as doves.

The first sentence clearly hints at the fact that Jesus knew that Israel's allegiance to God was nominal. The second sentence sounds like a proverbial expression, and it is Jesus' way of exhorting the disciples to use every way possible to protect themselves from those who would persecute them,

NILE CROCODILE

and is an echo of a saying of the rabbis, 'The holy and blessed God said to the Israelites, "Towards Me the Israelites are uncorrupt like doves, but towards the Gentiles they are as cunning as serpents."' We are back to the Garden of Eden.

When you read the Bible and come across a reference to 'snake', try to look at several translations; then you will have the best understanding of the way the writer wants you to think of the mention of the beast – a natural danger or symbolic of treachery or the embodiment of evil or just as a wild animal or symbolic of a Christian belief.

In God's discussions with the long-suffering Job, who has lost all his family and thinks he can no longer believe in God, God asks Job a mysterious question:

Can you pull in the Leviathan with a fishhook
or tie down his tongue with a rope?
Can you put a cord through his nose
or pierce his jaw with a hook?

JOB
CH. 41, VV. 1 2 & 7 8

...
Can you fill his hide with harpoons
or his head with fishing spears?
If you lay a hand on him,
you will remember the struggle and never do it again.

Professor Robert Alter in his commentary to his modern translation of Job, notes 'Leviathan is the fearsome primordial sea-monster subdued by the god of order in Canaanite mythology'. Most commentators think the Nile Crocodile *Crocodylus niloticus*, the second largest living reptile in the world, is the animal here. It is a beast the Israelites would certainly have got to know during their time in Egypt. The description clearly is referring to an animal that lives in water and would have been seen by fisherman who would have understood its strength and ferocity. The Psalmist (Psalm 104) and Isaiah (ch. 27) also write about this sea-monster, the latter calling it 'that wriggling, twisting dragon'. The Leviathan is surely a crocodile.

The main reason the Jews divided food into two classes, clean and unclean (see p. 87), was to preserve the sanctity of Israel as God's chosen people. One list in particular adds to our knowledge of Biblical reptiles:

LEVITICUS
CH. 11, VV. 29-30

Of the animals that move about on the ground,
these are unclean to you: ... any kind of great lizard,
the gecko, the monitor lizard, the wall lizard,
the skink and the chameleon.

NILE MONITOR

Various commentaries admit that the precise identity of some of these animals is uncertain. The 'great lizard' has been thought to be the Nile Crocodile *Crocodylus niloticus*, the second largest living reptile in the world; old mature males can grow up to 5.5 m (18 ft) long. It is widespread in fresh waters throughout Africa and formerly was found as far north as the Nile delta, so really could have been known to the Israelites when they were in Egypt. However, the *International Standard Bible Encyclopedia*, after careful argument considers the six reptiles in the list to be Thorny-tailed Lizard, monitor, gecko, true lizard, skink, and Chameleon; *The McClintock and Strong Biblical Cyclopedia* has a similar list of species but does not agree with which Hebrew word goes with which lizard! These species are the likely six: Rough-tailed Rock or Thorny-tailed Agama *Stellagama stellio* (formerly *Lacerta stellio*), found throughout the Middle East and common in farms and gardens; the Desert Monitor *Varanus griseus*, 2 m (6 ft) long, and found widely in the Middle East, including Palestine, or the equally long Nile Monitor *Varanus niloticus*, a native of Egypt; it would be good to think that the 'true lizard' is the Be'er Shaeva Fringe-fingered Lizard *Acanthodactylus beershebensis* which is found in the world only in south-central Israel in the Negev desert, but a small wall lizard is much more likely, such as the Lebanon Lizard *Phoenicolacerta laevis,* which is common; the skink may be the Ocellated Skink *Chalcides ocellatus,* one of many species found in sandy places, or the Sandfish Skink *Scincus scincus*; the Mediterranean Gecko *Hemidactylus turcicus* or the Fan-footed Gecko *Ptyodactylus gecko* are possibly the geckos; finally there is no doubt that the Common or Mediterranean Chameleon *Chamaeleo chamaeleon* is part of the list.

There are over 20 species of lizard recorded in Palestine, so what is a wonder is the fact that centuries before the discipline of zoology and the scientific naming of species, the Israelites had noticed the differences in the lizards and had names for the most obvious ones. The Bible has two particular references to lizards, in Leviticus, and the writer of Proverbs who counted 'four things on earth are small', one of which was 'a lizard [which] can be caught with the hand, yet it is found in king's palaces' (Proverbs ch. 30, v. 28). In warm countries like Palestine, geckos are small, are familiar sights indoors, clinging to ceiling or wall by means of their specially adapted toes that are covered in tiny hairs, which enable them even to walk up a window. They are most likely the lizards the writer had seen, even in a palace.

For all their importance in the 'unclean food' list, there are no other lizard references in modern translations, although the *Authorised Version* tried hard with the Hebrew word 'tannin' which modern

Mediterranean Chameleon

WALL LIZARDS

scholars are still struggling with. There we read that 'dragons and owls will honour' Isaiah, Job thinks he is 'a brother to dragons' and Jeremiah believes Jerusalem will become 'a den of dragons'. Modern translations refer to 'jackals' every time because the context suits that prowling animal, and we do not have the superstitious belief in dragons that the 17th century had.

Two other reptiles are worthy of note. The second great plague which was inflicted on the Egyptians was a plague of frogs, most particularly the Egyptian or Dotted frog *Rana punctata,* so named because its brown colour is covered with green spots:

EXODUS
CH. 8, VV. 6-11

Aaron stretched out his hand over the waters of Egypt, and the frogs came up and covered the land. ... Pharaoh summoned Moses and Aaron and said, "Pray to the Lord to take the frogs away from me and my people, and I will let you go to offer sacrifices to the Lord". ... Moses replied, "It will be as you say, so that you know there is no-one like the Lord our God. The frogs will leave you and your houses."

But Pharaoh did not keep his word until eight more plagues had struck. The plagues were to show that the Hebrews' god was a living God, and the Egyptian gods were worthless. The frog plague was a judgement against Heqet, the Egyptian frog-headed goddess of birth. The writer of Revelation (ch. 16, v. 13) names frogs in one of his visions as 'evil spirits' who will lead the people to support the cause of evil; so frogs are an image in keeping with their being 'unclean'.

Animals and birds, they are mostly comfortable with, but reptiles – snakes and crocodiles – no!

Two other animals are worth mentioning here, although neither is a reptile nor a mammal. The Snail *Helix* sp. is mentioned only once in the scriptures when the Psalmist prays that God will support the righteous and judge the unjust rulers, especially with regard to the latter's treatment of the poor and powerless:

PSALM
58, v. 6-8

*Break the teeth in their mouths, O God;
tear out, O Lord, the fangs of the lions!
Let them vanish like water that flows away;
when they draw the bow, let their arrows be blunted.
Like a snail [or slug] melting away as it moves along,
like a stillborn child, may they not see the sun.*

This is a vivid image from real life as the writer must have seen it: the snail's trail on the stony ground does melt away as it dries in the hot sun.

There is only one reference to the leech, the enigmatic,

> ➷ PROVERBS
> CH. 30, V. 31
>
> *The leech has two daughters.*
> *"Give! Give!" they cry.* ➷

Leeches are segmented, bloodsucking worms, of which the Medicinal Leech *Hirudo medicinalis* is the best known, and is found in Israel. It introduces the writer's warning that we should never feel that we are dissatisfied and need more. In the original Hebrew there is no word for 'they cry'. So the statement sounds much more dramatic if we read that the leech has two daughters, both named Give. Many people today want more and more, and are slow to offer help unless there is a positive reply to the request, "What are you going to give me if I do it?" Businesses want more profit, the jealous want what you have, the 'Romeo' wants another girl, and the alcoholic wants another drink – which neatly lets the snake, where we began, have the last word too:

> ➷ PROVERBS
> CH. 23, VV. 31-32
>
> *Do not gaze at wine when it is red,*
> *when it goes down smoothly!*
> *In the end it bites like a snake*
> *and poisons like a viper.* ➷

It is widely believed that James, the brother of Jesus, was the author of the letter in the New Testament bearing his name. He wrote it in the early 60s AD for Jewish Christians of 'the twelve tribes scattered among the nations'. He spends much of chapter 3 exhorting his readers to 'tame their tongues', to make sure that they use it to praise rather than to curse. He emphasises this by saying:

> ➷ JAMES
> CH. 3, VV. 7-8
>
> *All kinds of animals, birds, reptiles and creatures*
> *of the sea are being tamed and have been tamed by man,*
> *but no man can tame the tongue. It is a restless evil,*
> *full of deadly poison.* ➷

Of those 'creatures of the sea', fish are mentioned nearly 70 times evenly throughout the Bible. Sadly for us who have got used to all creatures having a name, not one species is identified, not even by King Solomon whom we are told 'taught about animals and birds, reptiles and fish' (1 Kings, ch. 4, v. 33). The disciples Simon Peter and his brother Andrew were fisherman before Jesus called them and said "I will make you fishers of men" (St Matthew, ch. 4, v. 19), and several stories about Jesus mention fish – for example, the boy with five loaves and two fish which miraculously fed the crowd, the amazing catch of 153 fish, and Jesus' appearance to the disciples after the Resurrection and being given a piece of grilled fish (St Matthew, ch. 14, St John, ch. 21 and St Luke, ch. 24). The nearest we get is the popular name of St Peter's Fish being given to the commonest species caught in the Sea of Galilee. It is a member of the freshwater family of *Tilapia*. The common name comes from the connection with the conversation Jesus had with Peter about taxes, because the temple tax collectors had just arrived at the house in Capernaum:

ST MATTHEW
CH. 17, VV. 25-27

"What do you think, Simon?" he asked. "From whom do the kings of the earth collect duty and taxes – from their sons or from others?"
"From others," Peter answered.
"Then the sons are exempt," Jesus said to him. "But so that we may not offend them, go to the lake and throw out your line. Take the first fish you catch; open its mouth and you will find a four drachma coin. Take it and give it to them for my tax and yours."

Four drachma was the amount of annual temple tax for two men, about two days' wages each. Jesus is implying that the disciples belong to God's royal household but disbelieving Jews did not.

We note interestingly that there are no insects in James' list of animals being tamed. The Bible does, however, mention several – sometimes at length. Most of them mirror how we think today about these often tiny creatures of God's creation. I have heard, as I'm sure you have, several people say, "Why did God allow wasps and mosquitos and locusts to go two by two into Noah's Ark? They are dreadful pests!" This chapter is not the place to discuss the environmental benefits of these species, but we will find out shortly what The Bible says about them.

Bee-keeping – apiculture – was practised by the ancient Egyptians, and archaeologists discovered hives at a dig at Rehov in the Jordan valley dating back to 900 BC. It is hard to believe that James did not know about hives of bees or colonies of wild bees. Bees and honey are mentioned so often in scripture, one of the earliest has become proverbial:

EXODUS
CH. 3, VV. 7-8

The Lord said, "I have indeed seen the misery of my people in Egypt … So I have come down to rescue them from the hand of the Egyptians and to bring them up out of that land into a good and spacious land, a land flowing with milk and honey"

From the *Authorized Version* onwards the description of a good country to live in has been said to be 'flowing with milk and honey', in other words, it lacks nothing in the way of nutritious and sweet food. That is where the Israelites were now living, in Canaan. Long after the Exodus, King Saul

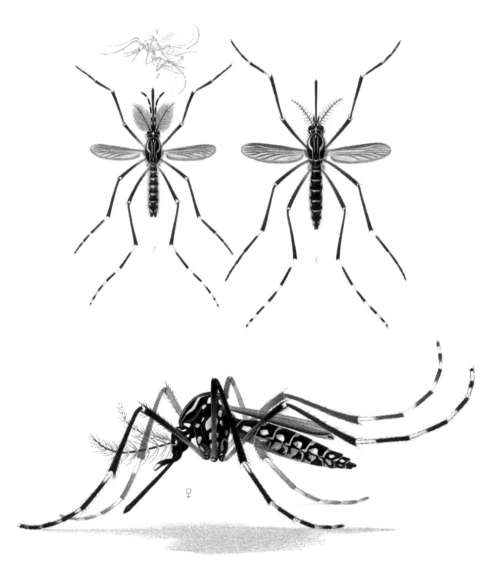

MOSQUITO

fought the Philistines, who were on the run, but before the Israelites pursued them Saul ordered his
men not to eat anything before evening came:

⤳ 1 SAMUEL
CH. 14, VV. 24-27

So none of the troops tasted food.
The entire army entered the woods, and there was honey on
the ground. When they went into the woods, they saw the
honey oozing out, yet no-one put his hand to his mouth,
because they feared the oath. ⤳

Saul learns that Jonathan has apparently disobeyed him and threatens to kill him, but is saved by the appeals of his own soldiers. 'Honey on the ground' sounds curious. Wild bees build a nest of honeycomb in the treetops, if that falls Jonathan could certainly poke his staff into it; bees do nest in a hole in the ground, such as Mining Bees and Bumble Bees; or it might be that the retreating Philistines dropped their food as they fled, which led Jonathan to say:

1 SAMUEL
CH. 14, V. 30

How much better it would have been if the men had eaten today some of the plunder they took from their enemies.

An even more extraordinary story concerns Samson. At his wedding feast he told a riddle to the 30 companions he had been given as was the custom among the Philistines:

"Tell us your riddle," they said. "Let's hear it."
He replied.
"Out of the eater, something to eat;
out of the strong, something sweet."
For three days they could not give the answer.

Samson's new wife pleads with him for the answer. On his way for his first visit to her he had killed a young lion. Later he had gone past it again, found a nest of bees in it and ate some of the honey, which in true Hebrew fashion meant that he was defiled, having touched a dead body. Eventually he gives in and tells her the answer. She of course tells her kinsmen, and:

JUDGES
CH. 14, VV. 1-20

Before sunset on the seventh day
[the deadline Samson had given the men to answer the riddle]
the men of the town said to him,
"What is sweeter than honey?
What is stronger than a lion?"
Samson said to them,
"If you had not ploughed with my heifer,
You would not have solved my riddle."

We usually think of Samson as a hero. Although he did give those who had solved the riddle the clothes he promised if they won, they were clothes he obtained by striking down and stripping 30 men of Ashkelon, one of the five principal cities of Philistine; this and his description of his wife, are two facts that reveal a very different character. Compared with the story we learn as a child that he lost his strength because of the evil doings of Delilah (who was in the pay of the rulers of the Philistines) which resulted in the 'hero's' death, he appears now as a *much* less appealing man, hardly a hero – he disobeyed his parents, failed to follow Hebrew law, rudely referred to his wife, angered the men of Ashkelon, was infatuated with Delilah, and above all 'he did not know that the Lord had left him' Judges, ch. 16, v. 20. And it all began with honey!

The Ant, one of over 10,000 known species in the family *Formicidae,* has come down to us in another proverb which I can remember being told me as a boy many years ago:

> PROVERBS
> CH. 6, VV. 6-8
>
> *Go to the ant you sluggard;*
> *consider its ways and be wise!*
> *It has no commander,*
> *no overseer or ruler,*
> *yet it stores its provisions in summer*
> *and gathers its food at harvest.*

Even after all those years I still find 'sluggard' a rare but interesting word, and was surprised and pleased that the *NIV* still uses it.

Ants certainly are not lazy. They are social insects that live in highly structured societies in nests they construct in trees, underground or in ground-level mounds. They do have a 'chief' known as a 'queen', but her role is only to produce thousands of eggs which are tended and guarded by 'workers', which are wingless females. Males have one function – to mate with the queen after which they die. The colony survives thanks to the perfect sharing of work, which is genetically built into each ant. They do not have to think about it; they just do it. Ants do collect food very busily. Some species 'milk' aphids and scale insects of their sweet secretion called 'honeydew'; various species of Leafcutter Ant *Atta* and *Acromyrmec* spp. cut leaves, take them to the nest and feed on the fungus which then grows on the cut leaves. Also in Proverbs we read:

> PROVERBS
> CH. 6, V. 25
>
> *Ants are creatures of little strength,*
> *yet they store up their food in the summer ...*

Several species do collect and store grain for the winter. But the writer was not as accurate with regard to his reference to ants' strength. For their size ants are very strong. Watch ants for any length of time, and you'll witness some remarkable feats of strength. Tiny ants, marching in lines, will haul food, grains of sand and even small pebbles back to their colonies. They can lift objects 50 times their own body weight. Their muscles – particularly neck muscles, because they lift the weight by holding it in their jaws – are thicker in proportion than those of larger animals or even humans. This ratio enables them to produce more force and carry larger objects. If we had muscles like ants we would be able to lift a family car over our heads!

Yet another saying we still hear today was originally said by the prophet Isaiah 3,000 years ago:

ISAIAH
CH. 51, VV. 7-8

Do not fear the reproach of men
or be terrified by their insults.
For the moth will eat them up like a garment;
the worm will devour them like wool.

ANT MILKING AN APHID

Locusts, Adults Flying and Wingless Immature below

Isaiah was quite right in observing that it is the moth's caterpillar that really does the damage, and its favourite food is wool. They can be a serious pest, and heated homes allow the moths to breed all year round. The adult moth does not feed and dies after it has mated. The moth was an image used by Jesus centuries later:

ST MATTHEW
CH. 6, VV. 19-21

Do not store up for yourselves treasures on earth, where moth and rust destroy, and where thieves break in and steal. But store up for yourselves treasures in heaven, where moth and rust do not destroy, and where thieves do not break in and steal. For where your treasure is, there your heart will be also.

The prophet was suggesting that troublesome men would be hurt, but Jesus was using the Clothes Moth *Tineola bisselliella* as an image to help people realise that material wealth does not last, and worse still, it means the person is 'worshipping' wealth not God.

There are detailed accounts of other insects that were – and still are – considered disagreeable or harmful. People today are often very distressed by the actions of wasps, clothes moths and locusts. The last named is arguably the most important insect in the Bible where it features strongly in many stories. Two species in particular occur in Palestine, the Desert Locust *Schistocerca gregaria*, widely distributed in North Africa, the Middle East and India, and the Migratory Locust *Locusta migratoria*, found from Africa to Australia.

The Israelites, led by Moses, were desperate to leave Egypt where they lived as slaves, but Pharaoh refused to let them go. Moses appeals to God who promises help, which comes in the form of dreadful natural disasters – the Nile stained red like blood, plagues of frogs, gnats, flies, diseased livestock, boils, hail, darkness, and penultimately, locusts:

EXODUS
CH. 10, VV. 13-15

So Moses stretched out his staff over Egypt, and the Lord made an east wind blow across the land all that day and all that night. By morning the wind had brought the locusts; they invaded all Egypt and settled down in every area of the country in great numbers. Never before had there been such a plague of locusts, nor will there ever be again. They covered all the ground until it was black. They devoured all that was left after the hail – everything growing in the fields and the fruit on the trees. Nothing green remained on tree or plant in all the land of Egypt.

The Psalmist years later recalled the ninth of the great plagues which afflicted the Egyptians before they allowed the Israelites to leave the country:

[God] spoke, and the locusts came,
grasshoppers without number;
they ate up every green thing in their land,
ate up the produce of their soil.

⤖ PSALM
105, vv. 24-25

The fact that the east wind blew the locusts to Egypt suggests they were Migratory Locusts. Locusts really do swarm like this, and farmers dread their arrival. The Israelites were released, but only after the last disaster, the death of every firstborn son, including the Pharaoh's.

Locusts are grasshoppers! There are ten Hebrew words used in scripture to signify locust. The insect is solitary in its first immature, flightless forms, until certain suitable conditions change it behaviourally to a gregarious insect, forming large bands and then swarms 'without number' as flying adults. Carried on the wind these swarms travel far. The writers of Exodus, the psalms and the prophet Nahum were right. Late in the 7th century BC he celebrated the fall of Ninevah the capital of Israel's enemy, the Assyrians, with these words:

You will be wiped out like crops eaten up by locusts.
You multiplied like locusts! You produced more merchants
than there are stars in the sky! But now they are gone,
like locusts that spread their wings and fly away.
Your officials are like a swarm of locusts that stay in
the walls on a cold day. But when the sun comes out,
they fly away, and no one knows where they have gone!

⤖ NAHUM
CH. 3, vv. 15-17
(GOOD NEWS BIBLE)

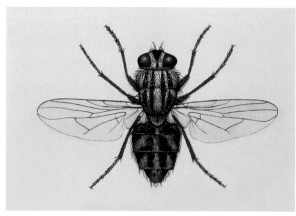

BARN FLY

A swarm might cover several square kilometres/miles, and number millions of insects, each devouring its own weight of food in a day, thousands of tons of precious crops, causing appalling devastation to valuable farmland such as happened in Pharaoh's Egypt. Swarms still occur today, as in Palestine in March to October 1915, and autumn 2004, which caused great hardship. Several organisations around the world today monitor the threat from swarms and endeavour to control the swarm by spreading insecticide over it from the air, or on the vegetation they are heading towards.

In the New Testament there are only three references to locusts. They are mentioned in St Matthew and St Mark with regard to John the Baptist, and in Revelation in a typical description of the horror they bring.

By the Mosaic law locusts were reckoned to be 'clean,' so that they could lawfully be eaten. Locusts are prepared as food in various ways. Sometimes they are pounded, then mixed with flour and water, and baked into cakes; 'sometimes boiled, roasted, or stewed in butter, and then eaten.' They were eaten in a preserved state by the ancient Assyrians as is shown in a carving nearly 3,000 years old that depicts a waiter bringing several dozen fixed on skewers to a feast – locust kebabs!

But no-one would wish to eat wasps! In many warm and tropical lands people who have a meal out of doors are often pestered by wasps or even its larger relative, the Hornet *Vespa* spp. It is recorded early on the story of the Jews' entry into the Promised Land. The Hebrew word 'tsir'ah' is used, meaning 'stinging', as when God says:

EXODUS
CH. 23, VV. 27-28

I will send my terror ahead of you and throw
confusion into every nation you encounter.
I will make all your enemies turn their backs and run.
I will send the hornet ahead of you to drive the Hivites,
Canaanites and Hittites out of your way.

The meaning of the Hebrew word translated here as 'hornet' is unclear. The *Good News Bible* admits it as a footnote; the New King James Version follows *NIV*'s lead. The word is also used by Joshua years later, and each time the Hornet is specified as a means of driving the Canaanites out from before the Israelites. Commentators have supposed that the word is used in a metaphorical sense as the symbol of some panic which would seize the people as a 'terror of God', the consternation with which God would fill the Canaanites. In Palestine the most likely species is the Oriental Hornet *Vespa orientalis*, differing from European Hornets *Vespa crabro* by being larger in size. They form a colony underground. It has been recorded they do attack human beings fiercely, and the furious attack of a swarm of Hornets drives cattle and horses to madness, and has even caused the death of the animals.

Other plagues that afflicted Egypt were of gnats and flies. The former became so numerous because 'All the dust throughout the land of Egypt became gnats' (although older translations say 'lice'), and:

EXODUS
CH. 8, VV. 17 & 25

Dense swarms of flies poured into Pharaoh's palace
and into the houses of his officials, and throughout
Egypt the land was ruined by the flies.

We don't know what species of gnat was involved. They are mostly tiny insects that do form large mating swarms, especially at dusk. They might even have been bloodsucking midges that carry diseases. The flies could well have been the Stable or Barn Fly *Stomoxys calcitrans,* as suggested by Dr John S. Marr in the television documentary 'Secrets of the Bible'. It is abundant wherever livestock are kept. It is even possible, likely in fact, that the flies or gnats were mosquitos, several small species of the family *Culicidae.* They are best known as the carrier of malaria by *Anopheles* spp. and breed in stagnant water; people living by the Nile were in prime mosquito habitat.

In the early days of the flight from Egypt the Israelites complained to Moses because of their lack of food. As we have read (p. 28), God provided manna, which they were to eat but not to save any overnight – they did and next morning 'it was full of maggots and began to smell' (Exodus ch. 16, v. 20). But a later word from God explained how they could collect twice as much on the sixth day of the week, bake it, boil it, and save whatever was left for the Sabbath next day, the day of rest, when they were not supposed to work. 'So they saved it until morning, as Moses commanded, and it did not stink or get maggots in it' (Exodus ch. 16, v. 24). It is fascinating what details are *not* in a story which we readers today would like to know – were the insects Mosquitos? – and what details *are* given – maggots in the Manna! An even more gruesome reference to such creatures, the larvae of some sort of fly, is in an anguished cry from Job:

> JOB *My body is clothed with worms and scabs,*
> CH. 7 V. 5 *my skin is broken and festering.*

Job was suffering so much he had no excuse for his wish to die. 'I despise my life' he shouts at God. Later in Job we read about God's long reply which describes so many creatures of His creation and Job is persuaded that he is wrong to be so angry.

Yet another disagreeable insect in the Bible is recorded solely in the story of David's troubled life with King Saul, who is looking for David in order to kill him. They meet and David says to Saul:

> 1 SAMUEL *The king of Israel has come out to look for a flea –*
> CH. 26, V. 20 *as one hunts a partridge in the mountains.*

David, at the same time that he is not afraid to picture himself as a tiny flea, suggests that King Saul is making a fool of himself by chasing him, an innocent man, with a great hunting party. The speech works, and Saul blesses David, says he will do great things and triumph, and they go their separate ways. There are about 2,000 species of flea worldwide, all wingless insects. They are external parasites, living by sucking the blood of animals and birds. David, as a soldier living rough, almost certainly knew at first hand the Human Flea *Pulex irritans*, well named because its bite does leave a very irritating, red mark; besides causing the itching, they are also the carriers of several diseases.

Two animals end this chapter which are not insects but are invertebrates as insects are (i.e. they have no backbone or spine). The first is the Spider, a member of the order of *Arachnids,* which have eight legs compared with the insect's six. There are over 40,000 species worldwide. Bildad, one of Job's friends who tries to comfort him, believes Job has upset God; that is why he is suffering:

⤝ JOB

CH. 8, VV. 13-15

Such is the destiny of all who forget God;
so perishes the hope of the godless.
What he trusts in is fragile;
what he relies on is a spider's web.
He leans on the web, but it gives way;
he clings to it, but it does not hold. ⤝

When we have brushed unexpectedly into a web we can appreciate its fragility. On the face of it, Bildad's illustration is good. Modern science however, has discovered that spider's silk has the tensile strength of steel. It is not as strong as steel but has a similar breaking point before it snaps. Many species of birds know this and use spiders' web to bind materials together in the construction of their nests – hummingbirds and the European Chaffinch are good examples.

SPIDER

When Ezekiel was 30, the age when a man could enter the priesthood, he heard a call from God:

 EZEKIEL
CH. 2, VV. 3-6

Son of man, I am sending you to the Israelites,
to a rebellious nation that has rebelled against me …
do not be afraid of them or their words. Do not be
afraid, though briers and thorns are all around you
and you live among scorpions.

Jesus often taught about prayer. He urged His followers to be bold or persistent in prayer, not just to pray on special occasions:

ST LUKE
CH. 11, VV. 9-12

So I say to you: Ask and it will be given to you; seek and
you will find; knock and the door will be opened to you.
For everyone who asks receives; he who seeks finds; and to
him who knocks, the door will be opened.
Which of you fathers, if your son asks for a fish, will give
him a snake instead? Or if he asks for an egg, will give
him a scorpion?

Both appeals name the Scorpion, an eight-legged creature in the same class of animals as spiders. There are over 1,000 known species throughout the world, mostly in warm climates. All are predatory, but only a few are poisonous; these are naturally the ones which humans are particularly aware of, as Moses reminded the Israelites:

DEUTERONOMY
CH. 8, V. 15

[God] led you through the vast and dreadful
desert, that thirsty and waterless land, with
its venomous snakes and scorpions.

The writer of Revelation was well aware of the Scorpion's sting in the tip of its tail when he stated in his vision that the godless people would suffer agony 'like that of the sting of the scorpion' (Revelation ch. 9, v. 5). All the references to the scorpion are when the writer or speaker wants to make sure that the audience is aware of the unpleasantness or weakness of a choice in which God does not feature.

It is a shame that the writers of the Bible told us only about insects which they think of as symbols of disaster, or produce good food, or are an illustration of a good way to live. There are no dragonflies or butterflies, two glories of God's insect creation. 'I have sought and seen butterflies in Israel from Ein Gedi to the top of Mount Hermon,' butterfly expert Jeffrey Zablow has said. 'The number of butterfly species is impressive, largely because of Israel's location and Israel's many diverse habitats.' Zablow has written about his trips to Israel extensively on his website, Winged Beauty. In Israel there are nearly 150 species of butterfly. Mt Hermon is the richest butterfly species site in Israel, with no fewer than 100 different species! Mt Hermon is the southernmost point of distribution for 30 of these 100 species, which do not fly to any other areas in Israel. I'm sure the prophets, psalmists, historians, Christ who walked the hills, and early Christians saw butterflies, but when they spoke or wrote their minds were on other things.

We must be grateful nevertheless, for the insects the Bible does record, reminding us of the sweetness of honey, the industry of the ant, and the wisdom of not storing up treasure on earth.

FURTHER READING

Alter, R. (2010) *The Wisdom Books – Job, Proverbs and Ecclesiastes*

Baly, D. (1959) *The Geography of the Bible*

Beer, E. (2007) *Flora and Fauna of the Bible* (mostly pictures)

Bouquet, A. C. (1953) *Everyday Life in New Testament Times*

Fish, H. D. (1998) *Animals of the Bible* (with Caldecott Medal winning illustrations by D. P. Lathrop)

Goodfellow, P. (2013) *Birds of the Bible – a guide for Bible readers and Birdwatchers*

Goodfellow, P. (2015) *Fauna and Flora of the Bible – a guide for Bible readers and Naturalists*

Hadoram, S. (1996) *The Birds of Israel*

Hareuveni, N. with Frenkley, H. (2nd edition, 1988) *Ecology in the Bible*

Hastings, J. (ed.) (1909 (10th impression, 1946) *Dictionary of the Bible*

Heaton, E. W. (1956) *Everyday Life in Old Testament Times*

Hovel, H. (1987) *Checklist of the Birds of Israel*

Josephus, F. (c.93/94 AD) *Antiquities of the Jews*

Moldenke, H. and Moldenke, A. (2005) *Plants of the Bible*

Mullarney, K. *et al.* (1999) *Collins Bird Guide*

Parmelee, A. (1960) *All the Birds of the Bible*

Paz, U. (1957) *The Birds of Israel*

Porter, R. *et al.* (2010) *Field Guide to the Birds of the Middle East, revised 2nd edition*

Reid, C. (1993) *Berlitz: Discover Israel*

Swenson, A. A. (1995) *Plants of the Bible and How to Grow Them*

Tristram, Rev. H. B. (1884) *Fauna and Flora of Palestine*

All Bible quotations are from *The NIV Study Bible: New International Version* (1998 edition) of the International Bible Society, unless otherwise stated.

Other Bible quotations about the plants and animals we have studied can be found at several websites such as International Standard Bible Encyclopedia, The Jewish Virtual Library, The McClintock and Strong Biblical Cyclopedia, ChristianAnswers. net, biblestudytools.com, the Holman Bible Dictionary, or by just raising the name via Google.

ACKNOWLEDGEMENTS

This book is the result of many years of reading and pondering Bible stories, and discussing them in devotional meetings with other believers, and preaching about them. Even after all that, I felt I still needed to have the book's content and interpretation looked at by other church members and I am particularly grateful to friends at Crownhill Methodist Church, Plymouth, who spent time reading and commenting on the manuscript. Their support has encouraged me greatly. I hope that what you read helps you better to understand what the writers had in mind when they named the plants and animals and birds.

Index